Medicinal Plants of India
An Encyclopaedia

Medicinal Plants of India
An Encyclopaedia

Ravindra Sharma

2024

Daya Publishing House®
A Division of
Astral International Pvt. Ltd.
New Delhi – 110 002

First Published, 2003

Reprinted, 2024

ISBN: 978-81-70353-04-1 (HB)

Disclaimer:

Every possible effort has been made to ensure that the information contained in this book is accurate at the time of going to press, and the publisher and author cannot accept responsibility for any errors or omissions, however caused. No responsibility for loss or damage occasioned to any person acting, or refraining from action, as a result of the material in this publication can be accepted by the editor, the publisher or the author. The Publisher is not associated with any product or vendor mentioned in the book. The contents of this work are intended to further general scientific research, understanding and discussion only. Readers should consult with a specialist where appropriate.

Every effort has been made to trace the owners of copyright material used in this book, if any. The author and the publisher will be grateful for any omission brought to their notice for acknowledgment in the future editions of the book.

Published by : **Daya Publishing House®**
A Division of
Astral International Pvt. Ltd.
– ISO 9001:2015 Certified Company –
4736/23, Ansari Road, Darya Ganj,
New Delhi-110 002
Ph. 011-43549197, 23278134
E-mail: info@astralint.com
Website: www.astralint.com

Printed at : Replika Press Pvt. Ltd.

डा. जगदीश कुमार रावत
निदेशक
Dr. J. K. Rawat
Director

भारतीय वन सर्वेक्षण
FOREST SURVEY OF INDIA
(भारत सरकार, पर्यावरण एवं वन मंत्रालय)
(Govt. of India, Ministry of Environment & Forests)
कौलागढ़ मार्ग, देहरादून-248195
Kaulagarh Road, Dehradun-248195

Foreword

Medicinal plants constitute a group of industrially important crops, which are of immense social, economic and commercial value. Plant-based drugs are being increasingly preferred in medical science today. The use of various parts of certain wildly growing plants to cure specific ailments has been in vogue in our indigenous medicine from ancient times. We are fortunate to have valuable information gathered by the native people through experience of generations in the form of folklore medicine. It is, however, unfortunate that full potential of these crops has not been scientifically explored. Therefore, in view of the importance of this group of plants and the need of the day, this book is a welcome contribution.

The present publication **"Medicinal Plants of India - An Encyclopaedia"** provides comprehensive and condensed information of an extensive and large number of medicinal plants grown and found in different agro-ecological zones of India. The publication describes, along with illustrations and photographs, nearly 600 species and 380 genera belonging to 172 families of the medicinal plants. The author has made special efforts to records vernacular (local) or trade names of these plants , which will be very helpful to the readers and users. The brief description of medicinal plants includes habit, vegetative features, floral characters and period of flowering (phenology). It is followed by information pertaining to local distribution and range in terms of elevation, topography, climate, vegetational composition, part used and chemical constituents. Beside this, the author has copiously incorporated medicinal uses along with traditional information. The author has also identified and listed certain threatened medicinal species and suggested possible *in situ* and *ex situ* conservation strategies.

The encyclopaedia will be extremely useful to forest managers, policy makers, researchers, scientists, cultivators, entrepreneurs, traditional healer (*vaidyas*) and other user groups. Presence of this document in libraries will help raise the awareness on medicinal plants as an important forest resource, and will ensure that medicinal plants are appropriately included in forest conservation and utilization programmes.

I am sure that the document will serve as a useful reference for important medicinal plants found in this part of the world. I congratulate Dr. Sharma over this remarkable achievement.

J.K.

(Dr. J. K. Rawat)

Date : 01-11-2002

Dehradun

Preface

Traditional systems of medicine ultimately show us that plants have provided mankind a large variety of patents drugs to alleviate suffering from diseases. However, drugs used in traditional systems of medicine are all crude drugs in their natural state or as their preparations. Until 19th century, even western medicine (allopathy) depends largely on crude drugs. During 20th century the progress in chemical techniques and with the growth of Pharmaceutical industry, crude drugs were replaced gradually by chemical (synthetic) drugs, however contending with various mechanism in human body to prevent their excessive actions, have a single mode of action and may cause severe adverse reactions when given in large doses or over a long period of time or some time differ with individuals. Thus synthetic drugs can produce remarkable life saving results in acute diseases but cannot be used often in the treatment of chronic diseases. A synthetic drug or a single chemical compound isolated from plants at times are also known to cause various estrogenic or physician-induced hazards.

On the contrary, a herb or crude drug as used in a traditional systems of medicine is a complex potpourri of compounds, some beneficial, some harmful, some vitamins and some even toxic – but all integrated under a certain (natural) rule to make a crude function in the same way as a single chemical agent. This crude drug, thus acts as a single chemical agent without any or much side effects unlike synthetic or chemically isolated single compound from the plants. Thus, a crude drug have substances amalgamated in a fashion where one chemical counter balance the undesirable effects of the other, ultimately to give only a beneficial effects. It is the reason why generally natural medicines induce fewer side effects than conventional western drugs.

The author have attempted to organize this book in such a way that it will be useful to everyone interested in the green health therapy. A number of valuable medicinal herbs are threatened with extinction, because of unsustainable over-exploitation and habitat destruction. The author endeavour here is to make people aware of the potential of medicinal herbs from all angles so that these life-saving plants can be protected from total destruction.

There is a need to identify the natural plant wealth and commercially viable species in each agro-climatic zone; promote industries to produce phytochemicals; standardize the agro-techniques for cultivation and production of these plants particularly in forests, wasteland as an additional income generating activity for evolving new herbal products and their production technology. In this book more 380 genera are discussed covering nearly 600 species. An integrated and comprehensive account of all related aspects of the subject such as nomenclature, description, distribution, mode of propagation, phenology, part used, active constitute of a large selection of plants has been incorporated.

I would like to inform you that for treating any ailments, the plants drug should be administered only with the prescription of a registered medical practitioners. This publication draws attention to the problems involved in a sensitive way and reflects well the new attitudes of contemporary ethnobotany which seeks both to protect the rights of indigenous peoples and to conserve their botanical heritage. At the same time, however, this does not exclude the sustainable use of these plants. This publication would serve as a practical guide and would be highly useful to the students, Vaidya's, Hakims, practitioners and other user groups, those working with medicinal plants.

It is my pleasure to acknowledge the help which I received from many publications for completing the book. Much of the information was gleaned from the libraries of the Botanical Survey of India (Northern Circle), Forest Research Institute (FRI), Dehradun, Indira Gandhi National Forest Academy (IGNFA), Dehradun, Indian Agricultural Research Institute, New Delhi.

I am also grateful to numerous inhabitants who had been kind enough to provide local information about the medicinal plants. Sincere thanks are also due to my wife Smt. (Dr.) Ranjana Sharma who has been a source of inspiration to pursue the work in all the circumstances. Words fail to express sense of indebtedness to my parents Sri Mam Chand Sharma (Civil Engineer) and Smt. Vimla Devi, who imbibed in my mind the aesthetics of the Nature.

I am also thankful to Mr. Anil Mittal, The Publisher, Daya Publishing House, deserves special appreciations for his painstaking and enthusiastic approach to complete this strenuous task.

While every efforts has been made to avoid mistakes, there could be still some errors. We would be grateful if the readers could kindly point them out, so that subsequent editions could be improved.

Dehradun, Uttaranchal

Dr. Ravindra Sharma

Contents

Introduction

India enjoys the privilege of having time tested traditional systems of medicines based on the natural products. Plants based products have been in use for medicinal, therapeutic or other purposes right from the drawn of history. The traditional remedies of the ancient world were all based on natural products, e.g. morphine from opium for use as narcotic and analgesic, latex exuded from the poppy seed for gout, cocaine from cocoa-leaf as potential local anesthetic, ergot for obstetric use, castor oil, senna etc. as laxatives, turmeric as an antiseptic, garlic for blood and heart remedies to mention just a few. The traditional Indian system of medicine, namely Ayurveda which involves dispensing of herbal and plant products in various forms such as powders, extracts, decoctions etc., dates back to the vedic period, when the first mention of diseases and drugs is found in the Rigveda and Yajurveda (*i.e.* around 2000 B.C.) and the earliest comprehensive description of Ayurveda is available in the Atharvaveda (*i.e.* 1600-1000 B.C), which contains inter alia descriptions of human anatomy, rudiments of classifications of diseases and reference to herbal medicine.

Recent past has witnessed an upsurge in the popularity of the herbal medicines. In the developing countries, about 80% of the people depend upon the traditional system of medicine and 95% of the industrial need of this is met through indiscriminate collection from wild so much so that over half a million tones of dry raw material is indiscriminately collection of plants has led to considerable genetic erosion and loss of biodiversity including plants of medicinal value.

Among endangered plant species, medicinal herbs account for almost 1/3 of the species mentioned in the red data book. In the absence of organized cultivation, 14 species have become endangered or rare and 35 species are vulnerable.

Medicinal plants as a group comprise approximately 8000 species and account for around 50% of all the higher flowering plant species of India. India possesses almost 8% of the estimated biodiversity of the world with around 0.126% million species. India is one of the 12 mega biodiversity centers with 2 hot-spots of biodiversity in Western Ghat and North-Eastern Region. There are about 400 families in world of the flowering plants, atleast 315 are represented in India

Conservation of biological diversity, particularly medicinal and aromatic plants is one of the 5 top priority environmental problems facing the globe. This is more true in case of developing countries located in tropical/subtropical regions of the world as they are the store-house for much of the biodiversity of relevance to human needs. To circumvent this situation, genetic erosion needs to be prevented through restoration of habitats and organized gene banks both *ex situ* and *in situ*, protected area networks, herbal gardens, field gene banks etc. The increasing the demand of phyto-medicines calls for concerted efforts for domestication and cultivation of genetic resources of medicinal plants on large scale through herbal gardens.

The development of systematic pharmacopoeias dates back to 3000 BC, when the Chinese were already using over 350 herbal remedies. Ayurveda, a system of herbal medicine in India, Sri Lanka and South-east Asia has more than 8000 plants remedies and using around 35,000-70,000 plant species.

China has demonstrated the best use of traditional medicine in providing the health care. China has pharmacologically validated and improved many traditional herbal medicines and eventually use them in formal health care system.

The phyto-medicines are also being increasingly used in Western Europe. Recently the US Government has established the "Office to alternative Medicine" at the National Institute of Health at Bethesda and its support to alternative medicine includes basic and applied research in traditional systems of medicines such as Chinese, Ayurvedic etc. with a view to assess the possible integration of effective treatments with modern medicines. Of about 240 thousand known species of higher plants, Chinese, Indian Arabian and other traditional systems of medicines make extensive use of about 5000 plants. Medicinal plants continue to provide health security to rural population.

Medicinal Plants Demand and Supply

In India, there are 7843 licensed pharmacies of Indian System of Medicine in addition to 857 of Homeopathy and a number of unlicensed small scale processing unit engaged in the manufacturing of the medicines to meet the requirement of 4.65 lakhs registered practitioners of ISM & H and other users. Although about 8000 species of plants are estimated to be used in human and animal health care and over 10,000 herbal drug formulations have been recorded in codified medical texts of Ayurveda, the pharmaceutical industries are largely based on about 400 plant species.

According to 1994 UNDP report, the annual value of medicinal plants derived from developing countries is approximately 32 billion US dollars. There are 47 major modern pharmaceutical plant-based drugs already in the world market and the predicted 328 drugs, yet to be discovered, have a market potential of 147 billion US dollars. The sale in the very first year of the anticancer drug, taxol from *Taxus spp.* has been more than 2000 billion US dollars. Phyto-remedies and health-food over-the-countries sales in USA is expected to touch 2 billion US dollar according to American Botanical Council.

Unsustainable ways of harvesting and unrestricted marketing have led to the reduction in population of some of the high demand medicinal plants leading to sudden escalation in prices of these crude drugs in the market.

Marketing and Trade of Medicinal Plants

The value of global trade of the medicinal plant products has been put over US$ 75 billion per year and is growing @ 12.5% annually. Of the total value of trade, about 20 billion US dollar account for over the counter (OTC) drugs, US$ 25 billion for prescription drugs and remaining US$ 30 billion for the nutritional supplements.

Bulk of the raw material (90%) is produced in Asia, Africa and Latin America and some in Europe and USA (10%). About 60% of the total material is imported and processed in USA, Canada, UK, Australia, Germany, France, Italy, Switzerland and Japan and about 50% of that is used there and the rest is exported to the raw material producing countries to be sold at high rates. About 90% of marketed material is collected from the wild resources and there is some cultivated in China, India, USA, Germany, France, Italy and Eastern Europe.

The trade in medicinal plants in India is estimated to the tune of Rs 675 crores per year. Of India's total turnover of Rs. 3,100 crores of Ayurvedic and herbal products, major OTC products contribute around Rs. 1,700 crores, Ayurvedic ethical formulations constitute around Rs. 850 crores and Ayurvedic classical formulations constitute remaining Rs. 550 crores. Indian export has steadily grown at the rate of 65%, since 1991-92 and grew to Rs. 215 crores in 2000-01 from Rs.130 crores in 1991-92. It is estimated that the global market for herbal drugs is nearly Rs. 800 crores per year. The export of medicinal plant from India appears to be growing faster.

Conservative estimates put the economic value of medicinal plants related trade over US$ 90 billion. Demand and trade in medicinal plant species globally indicates a steep upward trend and World trade in medicinal plant and related products is expected to rise to US$ 5 trillion by 2050 A.D.

Conservation of Medicinal Plants: Strategies & Priorities

Several national and international agencies have formulated appropriate policies and strategies for the conservation of medicinal plants. The world conservation strategy (IUCN, UNEP & WWF, 1980) defines conservations as "the management of human use of the biodiversity so that it may yield the greatest sustainable benefit to present generation while maintaining its definitions invokes two complementary components "conservation" and "sustainability". The primary goals of biodiversity conservation as envisaged in the World Conservation Strategy can be summarized as follows:

(1) Maintenance of essential ecological processes and life support systems on which human survival and economic activities depend,

(2) Preservation of species and genetic diversity and

(3) Sustainable use of species and ecosystems which support millions of rural communities as well as major industries.

Medicinal plants are potential renewable natural resources. Therefore, the conservation and sustainable utilization of medicinal plants must necessarily involve a long term, integrated, scientifically oriented action programme. This should involve the pertinent aspects of protection, preservation, maintenance, exploitation, conservation and sustainable utilization. A holistic and systematic approach envisaging interaction between social, economic and ecological systems will be a more desirable one. The most widely accepted scientific technologies of biodiversity conservation are the *in-situ* and *ex-situ* methods.

In-Situ Conservation

In has been established that the best and cost effective way of protecting the existing biological and genetic diversity is the *'in-situ'* or on the site conservation wherein a wild species or stock of a biological community is protected and preserved in its natural habitat. The prospect of such a 'ecocentric', rather than a species centered approach is that it should prevent species from becoming endangered by human activities and reduce the need for human intervening to prevent premature extinctions.

Ex-Situ Conservation

Conservation of medicinal plants can be accomplished by the *ex-situ* i.e. outside natural habitat by cultivating and maintaining plants in biotic gardens, parks, other suitable sites, and through long term preservation of plant propagules in gene banks (seed bank, pollen bank, DNA libraries, etc.) and in plant tissue culture repositories and by cryopreservation.

The medicinal and aromatic plants should be domesticated and brought under cultivation to maintain constant supply of quality materials and thus reduce the pressure on the wild populations. The process of domestication of a species involves charàcterizations of its reproductive biology to decide on the method of propagation-seed or propagule and definition of the area in which to cultivate based on the soil and climate characteristic of the plant's natural habitat. The genetic resources of the species are screened for its adaptability and yield and quality of the material that will make the end product, to identify suitable genotypes for cultivation. Soil type to be used, irrigation and fertilizer amounts and application schedule and sowing and harvesting times are standardized. Disease and pest problems of standing crop and harvesting material are solved to keep the product safe for consumption. Initiation of plants breed programme will ensure that in future the crop will be high yielding, resistant to pest and domestic quality.

Cultivation and processing of material harvested from medicinal and neutraceutical (Supplements) plants, already identified useful for restoration and protection of human health, can provide much needed avenues of self-employment to educated unemployed in villages and small towns. Prospection of plant species for new drugs and other products and genetic engineering for certain crop plants for the production of raw material for the secondary metabolites for pharmaceutical industry are expected to further open such employment opportunities. To make the above developments possible, it is of immediate paramount importance to inventories evaluate and conserve plant species in nature and to *ex-situ* conserve the genetic resources domesticate and cultivate species whose medicinal and industrial importance is already established.

Abelmoschus moschatus Medik.

Family : Malvaceae
Trade/Common Name : Mushkdana

A hardy herb. A herb or undershrub, hirsute with simple or stellate hairs upto 2.5 m high. Leaves angular or 3-7 palmatilobed upper leaves narrower, lobes linear-lanceolate to oblong-ovate, serrate to dentate; stipules linear to filiform, simple, hairy. Epicalyx bracteolates 6-10, free linear persistent. Flowers solitary, axillary, yellow in colour and blooming in the month of July to continue up to September.

Abelmoschus moschatus

It is grown in grassy slopes, forest edges of sub-montane Himalaya and hotter parts of India. It is propagated by seeds.

Seeds yields an essential oil known as ambrette oil. It is used in perfumery, cosmatic, and agarbathi industries. Besides, it ia also used for flavouring chewing tobacco, tea products and in various medicines.

Abies spectabilis (D.Don.) Spach

Syn. *A. Webbiana* Linn.

Family : Pinaceae
Trade/Common Name : Himalayan Silver fir

A lofty, hardy, coniferous evergreen tree of great height. Leaves flattened, linear, spirally arranged around the branches; tip

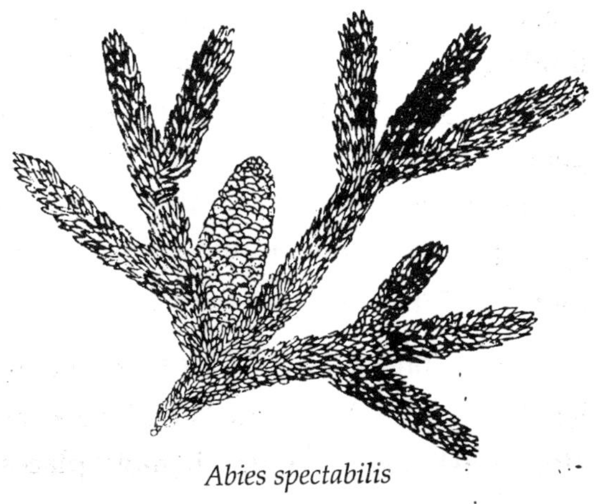

Abies spectabilis

notched, upper surface dark green, shining; lower with a silvery white line on each side of the midrib with incurved margins. Flowering appears in April-May. Cones ripen during September-October. It is propagated by seeds but sometimes grafting is possible.

It is found from Afghanistan to Bhutan, common in Uttaranchal at elevation 2800-4000 m spreading over large areas in the upper tree zone. It is propagated by seeds.

Leaves and needles are medicinally important and contains essential oil and an alkaloids (0.16% dry wt.) It is also reported antiflatulent, antispasmodic, appetizer, astringent, carminative, digestive, diuretic, expectorant, febrifuge, stomachic and tonic.

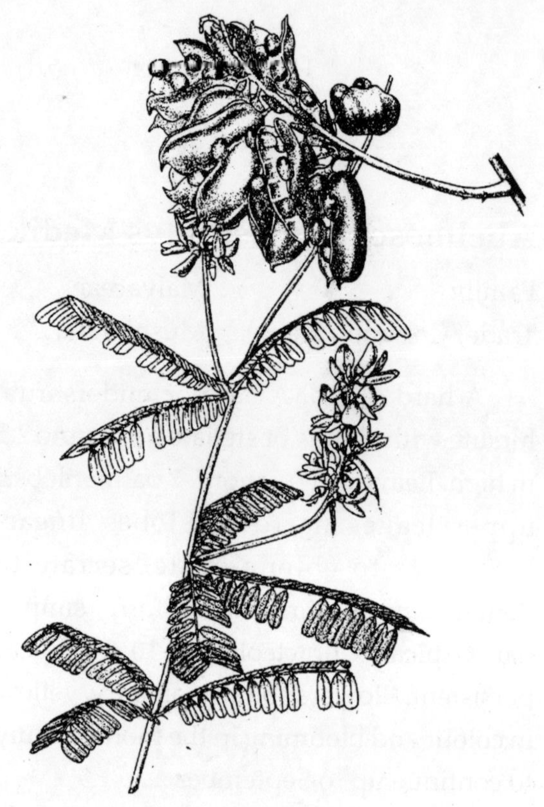

Abrus precatorius

Abrus precatorius Linn.

Family : Fabaceae

Trade/Common Name : Indian Liquorices

A much branched, perennial climber with woody stem. Leaves are paripinnate; leaflets opposite, deciduous, oblong, ligulate, membranous, glabrous or thinly silky beneath. Seeds ovoid or sub-globose, generally two third scarlet or white and the rest jet black giving appearance of a black eye. Flowering and fruiting appear in the month of September to January.

It is found in South China, the pacific Island, Tropical Africa, the West Indies and almost all tropical countries in moist places. In India, It occurs commonly near the ravines of miscellaneous forests, waysides, grassy localities up to 1000 m altitude and throughout warmer regions of country. It is propagated by Seeds and can also be propagated by cutting of firm shoots in sand under glass.

Seeds and Roots are medicinally important. Seeds contain abrin, abrine and abarnin while roots contain precol, abrol and two alkaloids abraime & precasine. Seeds are purgative, emetic, tonic, aphrodisiac in nature. Roots are emetic, alexiteric. Locally, the leaf juice mixed with oil is applied on painful swellings of the body. The root is used as a substitute of liquorice.

Abutilon indicum (L.) Sw.

Family : Malvaceae
Trade/Common Name : Country Mallow

A erect herb with cylindrical, branched, solid, pubescent, green stem. Leaves cauline, ramal, alternate, simple, petiolate, deltoid, serrate, acute, slightly hairy, and reticulate. Flowers pedicellate, golden yellow in colour, arises on solitary axillary and blooming in the month of August to continue upto October.

It is found as weed in various areas of India. It is propagated by seeds or stem cutting.

It possesses diuretic, demulcent and laxative properties. It is also used to treat fever.

Acacia catechu (L.f.) Willd.

Family : Fabaceae
Trade/Common Name : Catechu Tree

A moderate sized tree, 10-13 m. Its grayish brown bark, exfoliating in long, narrow strips, brown or red inside. Leaves are bipinnate, armed with stipulary thorns. Flowers are small, white or pale yellow in axilliary spikes and blooming in month of May-June but fruiting upto October.

It is found commonly dry exposed miscellaneous forests, in India from Indus

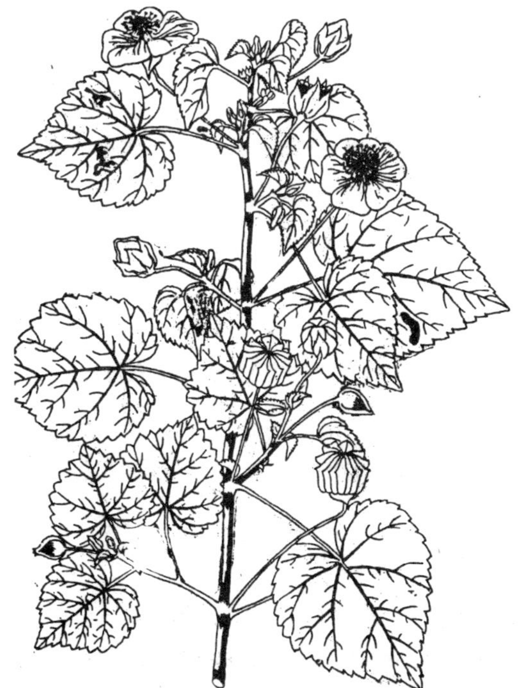

Abutilon indicum

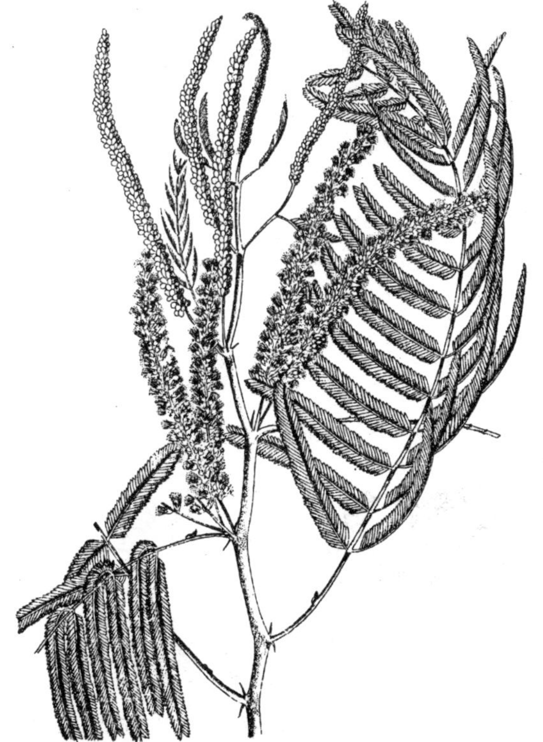

Acacia catechu

eastwards to Assam and throughout Peninsula, particularly in dry regions. It is propagated by seeds or cuttings.

Heart wood, extract (Katha) and bark are medicinally important and contains catechin, catechutanic acid and tannin. Bark is used as astringent, digestive and useful in cough, diarrhea, bronchial affections cancer, piles, sore throat, and ulceration and obstinate skin diseases. Locally, the bark decoction is taken in the morning and evening for three days in stomache.

Acacia nilotica (L.) Willd ex Delile

Syn. A. arabica (Lam.) Willd.

Family : Fabaceae
Trade/Common Name : Indian Babul

A moderate sized tree with a large spreading crown, bark dark-brown almost black, much fissured, spines long, straight white, sharply pointed. Leaves 2.5-5.0 cm long, rachis and pinnate downy pinnate 3-6 pairs. Flowers yellow, fragrant in axilliary, globose heads. Its flowering and fruiting appears in month of July-September to April-June.

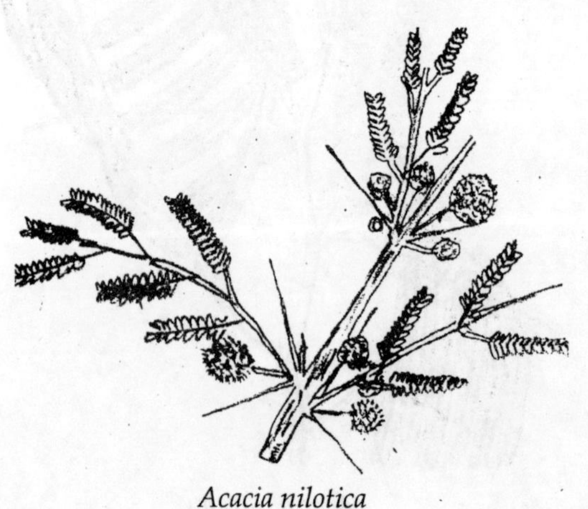

Acacia nilotica

It is found in throughout the warmers regions of India ascending to 800 m. It is propagated by seeds.

Heart wood, Gum, Pod and Bark are medicinally important and contains several polyphenols like catechin, epicatechin, dicatechin, quercetin and tannin. Various plant parts are used against hairfall, earache, syphilis, cholera, dysentery, leprosy and rinderpest.

Acampe papillosa Linn.

Family : Orchidaceae
Trade/Common Name : Nakuli

An epiphyte orchid with erect stem and densely leafy upto 50 cm in height. Leaves thick, recurved, long, fleshy. Flowers fragment, dark yellow spotted brown in colour but lips white.

It is native to Eastern Himalaya, Khasia hills and Assam.

Achillea millefolium Linn.

Family : Asteraceae
Trade/Common Name : Millfoil

A small, perennial tufted herb with a slander cropping rootstock throwing numerous roots and stolons with blunt succulent scale at each node. Leaves alternate, oblong-lanceolate. Flowers are

Achillea millefolium

numerous, small, white in corymbose, ovoid flat topped heads at the stems and branches and blooming in the month of July-September.

It is found in temperate and alpine areas from Pakistan to Uttar Pradesh, Temperate Eurasia 1800-3600 m in meadows. It is propagated by division of roots and through seeds.

Leaves and flower heads are medicinally important. Its infloresance yields volatile oil (0.25% dry basis) contains azulene (1-5%), achilloic acid, alkaoil-achilleine, a better principle, ivain and tannin and employed as a carminative, tonic, and aromatic stimulant.

Achyranthes aspera Linn.

Family : Amaranthaceae
Trade/Common Name : Prickles chaff

An erect herb about 0.5-2.0 m high with often a woody rootstock and much branched hairy or sub-orbicular, acute, obtuse or acuminate, sparingly hairy with a short petiole. Flowers are green or pinkish crowded in terminal spikes and blooming in month of April-August.

It is found throughout India upto 1300 m. As a weed, it is found in perennial crops, grasslands, waste places, shaded or unshaded areas. It is propagated by seeds.

The whole plant is medicinally important. Seed contains two saponins;

Achyranthes aspera

saponin A and saponin B. Saponin à-methyl ester, b-methyl ester, have also been found. The herb possess antidiabitic and antirheumatic properties. Seeds are used as inflammatory, acrdiotonic and diuretic and used to treat snake bites, hydrophobia and itchings.. Its powder is used in the treatment of bleeding piles. Locally, the decoction of the herb is diuretic and used in renal dropsy and generalized anasarca. Leaf juice is useful in stomachache, bowel compliments, piles and abscess. The extract of the root is used to treat menstrual disorders and dysentery.

Aconitum balfourii Stapf.

Family : Ranunculaceae
Trade/Common Name : Meetha bish

A biennial herb with tuberous roots. The flowers hooded and foliage delphinium-like and blooming in the month of July to continue upto September.

It occurs in shady slopes and forest edges of Uttaranchal Himalaya between 3000-4200 m altitude. It is propagated by divisions of rootstocks.

Roots are used against neuralgia and rheumatism. Tubers said to be toxic and narcotic used for medicinal purposes in fever and bonel complaints.

Aconitum deinorrhizum Stapf.

Family : Ranunculaceae
Trade/Common Name : Indian Aconite

A tall, glabrous herb with paired tuberous roots with dark brown colour in

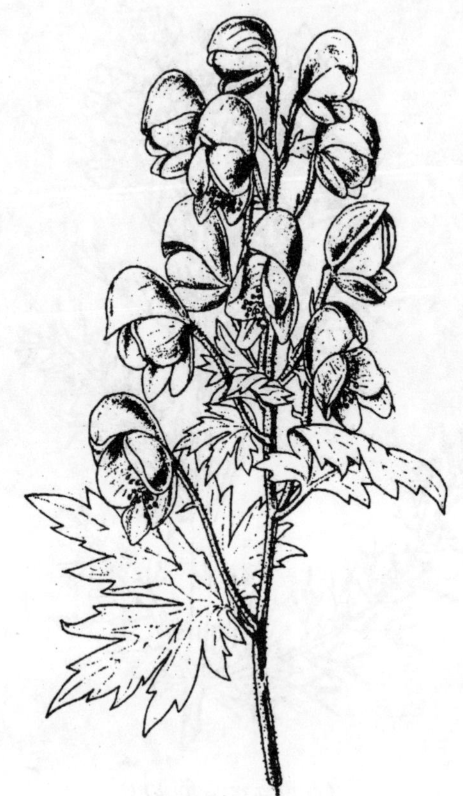

Aconitum balfourii

surface and coarsely wrinkled. Leaves are scattered, reniform or ovate reniform, 5-10 cm long, acute. Cauline leaves, sharply toothed, lower long petioled. Flowers are bright blue to greenish blue with purple veins in terminal racemes and blooming in the month of July-September.

It is found common in the grassy slopes of sub-alpine and Alpine zones of the Himalaya from Indus to Uttarachal at altitude between 2400-4500 m. It is propagated by divisions of rootstocks.

Roots are medicinally important and contains an alkaloids aconitine and pseudoaconitine (0.4%). Roots are toxic and affects both the heart and nervous system

Aconitum deinorrhizum

and also used in leprosy, cholera, rheumatism, as tonic in diarrhea and cough.

Aconitum falconeri Stapf.

Family : Ranunculaceae
Trade/Common Name : Vibh

An erect, biennial herb upto 1 m tall. Flowers hooded and foliage delphinium-like and blooming in the month of July to continue upto September.

It is native to sub-alpine and alpine zone of Himalaya in Uttaranchal between 3000-4000 in elevation. It is propagated by division of tuberous roots.

Aconitum falconeri

Leaves, flowers and tuberous roots are used in medicinal preparation. It possesses antidiabitic, antipyretic and diuretic properties. Ayurvedic preparations are used for curing skin diseases, rheumatism, neuralgia fever and diarrhoea.

Aconitum ferox Wall.

Family : Ranunculaceae
Trade/Common Name : Bikh

A perennial herb with tuberous, cyclindrical roots with greyish colour. Stem sub-erect to procumbent. Flowers are deep

Aconitum ferox

blue in terminal racemes and blooming in the month of July-August.

It is common, occasionally found in the open slopes of Sub-alpine and Alpine meadows of cold arid region of Sikkim Himalaya at altitude between 2100-4000 m. It is propagated by divisions of rootstocks.

It is used in making Ayurvedic preparations and is one of the important ingredients of Homeopathic orescriptions. It possesses alternative, diaphoretic, diuretic, expectorant, febrifuge and diaphoretic properties. It is useful in diabetes, debility, asthma, ear and nose discharges, leprosy, paralysis. Its powdered is administered orally against rheumatism, fever etc. The root is used as an antidote to hartal, believed to be a lethal poison of local origin.

Aconitum heterophyllum Wall. ex. Royle.

Family : Ranunculaceae
Trade/Common Name : Atish

An erect, leafy herb having biennial, paired, tuberous roots. Stems upto 1 m tall with broad, ovate or orbicular, cordate, lobed and toothed, shortly stalked or sessile, amplexicaule leaves. Flowers bright-blue to yellow greenish blue with, purple veins in long and blooming in the month of July-August.

Aconitum heterophyllum

It is found in grassy meadows of temperate to alpine zone of the Himalaya, from Indus to Uttaranchal at altitudes between 2500-4000 m. It is propagated by seeds and by divisions of roots.

The roots are medicinally important and contains non-toxic, amorphous alkaloids, atisine (0.4%), dehydroatisine, heteratisine, hetisine, aconitic acid, tannic acid, pectin and starch. It is used as antiarthritic agent, astringent, carninative, febrifuge, stomachache and aphrodisiac in the various indigenous system of medicines. Locally, its small doses mixed with tobacco for curing cold, cough and stomachache.

Aconitum rotundifolium Kar. & Kir.

Family : Ranunculaceae
Trade/Common Name : Cherkhan

A perennial herb with paired tuberous roots with dark brown colour in surface and coarsely wrinkled. Leaves are rhomboid, 5-7 lobed, each lobe 2-3 fid. Flowers are deep blue in terminal racemes and blooming in the month of July-September.

It is rare, gregarious in nature and found in the moist slopes of Alpine zones of cold arid region of Kashmir Himalaya at altitude between 3400-4500 m. It is propagated by divisions of rootstocks.

It exhibit weak anti-tumor activity and in Tibetian medicine, the water extract of root prescribed against severe headaches in small quantities.

Aconitum rotundifolium

Aconitum violaceum Jacq. ex Stapf.

Family : Ranunculaceae
Trade/Common Name : Patis

A perennial herb with tuberous, cylindrical roots with greyish colour. Stem sub-erect to procumbent. Flowers are deep blue in terminal racemes and blooming in the month of July-September.

It is common, occasionally found in the open slopes of Sub-alpine and Alpine meadows of cold arid region from Kashmir to Uttaranchal Himalaya at altitude between 3400-4500 m. It is propagated by divisions of rootstocks.

It used as a pleasant tonic. The roots are poisonous in anture and fresh root powder is taken with hot water is acse of fever due to bad cold in Kumaon Himalaya. Dried root pieces are fitted in tooth cavities against toothache. Locally, it is given to infants in the cold, cough and stomache ache by mixing root power with mother's milk.

Acorus calamus Linn.

Family : Araceae
Trade/Common Name : Sweet flag

An aeromatic semi-aquatic or marshy, perennial herb with indefinitely branched rhizomes, creeping in mud and have stout joints with large leaf scars. Rhizomes are cylindrical or somewhat compressed. Leaves are few distichously alternate, forming erect tufts at the extremities of rhizomes. Flowering appears in the month of April-July.

It is found Throughout India and Pakistan in damp marshy open places, prefer waterlogged conditions. It is propagated by cutting of rhizomes.

The dried rhizomes are medicinally important and contain a yellow bitter aromatic volatile oil (1.5-3.5%). It contains a-pinene, d-camphene, calamene, calamenol, calamenone and alcohols. The roots also contain a glucoside, 'Acorin', calamene, tannin, mucilage, starch and vitamin C. Powered roots are used as vermifuge, emetic, stomachic and carminative. Locally, its paste is applied

Acorus calamus

directly on the forehead during fever. Sometimes, in Kwashiorkor disease of children, small pieces of roots are tied around the neck.

Actaea spicata Linn.

Family : Ranunculaceae
Trade/Common Name : Herb Christopher

A hardy herbaceous perennial upto 60 cm tall with horizontal roots. Leaves ovate, lanceolate, three lobed and intire. Flowers are white in colour.

Actaea spicata

It is native to Britain and occurs in temperate range of Uttaranchal Himalaya between 2500-3500 m altitude. It is propagated by seeds and root cuttings.

Leaves, berries and roots are medicinally important. Root possesses violent emetic, nerve sedative and purgative properties. Dried powder of the herb is used as insecticides. Berries contain essential oil but toxic in nature.

Actiniopteris radiata (Swartz.) Link.

Syn. *A. australis*

Family : Actiniopteridaceae

Trade/Common Name : Sanjeevani

A evergreen fern with a spearding segmented fronds.

It is native of India and found throughout the Aravalli hills between 400-1000 m altitude. It is propagated by spores.

The whole is used in bronchitis and gynaecological disorders. The plant ash is administered in burns, wounds and scalds. The dried leaves are used in tuberculosis.

Adiantum capillus-veneris Linn.

Family : Adiantaceae

Trade/Common Name : Maiden Hair

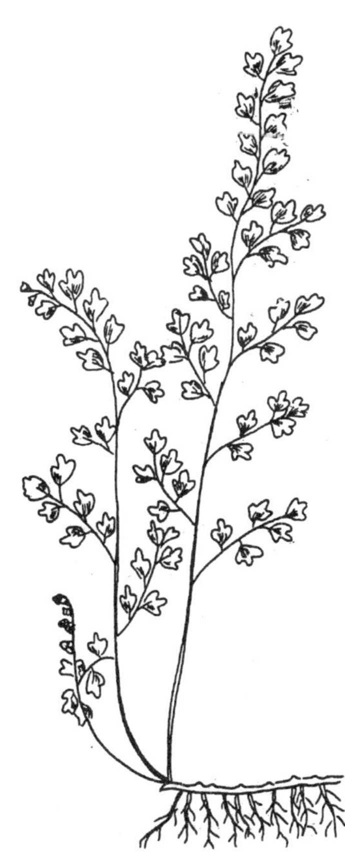

Adiantum capillus-veneris

A perennial herbaceous fern with long polished, black stripes. Fronds bipinnate, segments cuneate, upto 2 cm broad. Sori are borne at the rounded sinuses of the crenations.

It is found in mud walls, rocky and damp places in forest of Western Himalaya, extending upto Manipur. It also found in south India, Maharashtra, Punjab, Jharkhand and Bihar. It is propagated by spores.

It contains 3,4-Epoxyfilicane, 21-hydroxyadiantone, adiantone, astragaline, isoquercitrin and kaempferol. It possess demulcent, expectorant, diuretic, febrifuge properties and also used as hair tonic. The decoction of leaves is taken for acute bronchitis and fever. Locally, it is also taken along with tea in abnormal stoppage of the menses.

Adiantum venustum G. Don.

Family : Adiantaceae
Trade/Common Name : Maiden hair fern

A dark brown fern with fronds 3-4 times pinnate. Rachis slender, polished, naked, segments rigid, prominently veined and toothed, upper edge rounded, lower cuneate into the petiole. Sori 1-3, large, roundish, placed in a distinct hollow on the upper edge.

It is found in shady and moist sloped, edges of ravines, on stone walls and

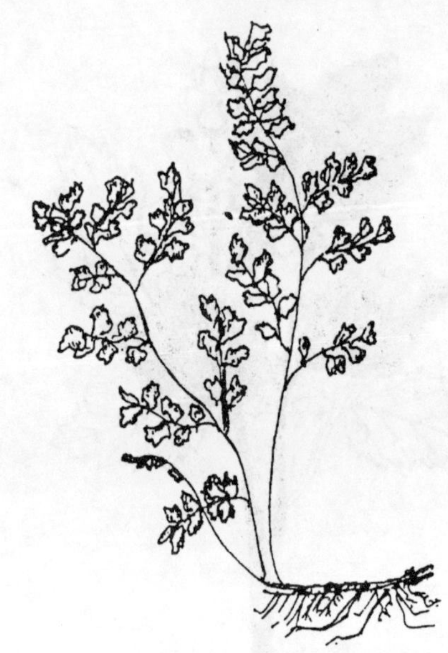

Adiantum venustum

comparable in most part of the Himalaya at altitudes of 2200- 2800 m.

The herb contains tannins and mucilage and used as astringent, febrifuge, tonic, and expectorant.

Adonis aestivalis Linn.

Family : Ranunculaceae
Trade/Common Name : Indian Adonis

An erect, perennial upto 50 cm tall. Leaves much divided. Flowers are scarlet with a dark-purple eye and appear at the ends of the branchlets.

It occurs in Western Himalaya. It is propagated by seeds and division of roots.

It is cardiac stimulant, abortifacient, acrid. Locally, its seeds used for colic.

Aegle marmelos (L.) Corr.

Family : Rutaceae
Trade/Common Name : Beal fruit

A medium-sized, deciduous tree, having strong axilliary thorns. Leaves trifoliate, stalked, alternate; leaflets oblanceolate, crenate, gland dotted; middle leaf is stalked and larger while the later one's are sessile. Flower greenish, fragrant and blooming in the month of May-June.

It is found in sub-montane regions and plains, almost throughout India. It is propagated by seeds.

The fruit contains marmalosin, bark contains coumarin, and umbelliferone; Leaves are having essential oil (0.6%), consisting of a and b-phellandrene. It is useful in chronic dysentery, diarrhea, appetite and digestion. Locally, the bark decoction is taken thrice a day for one week in tntermittent fever. The root of one of the ingredients of the 'Dasamool' of Ayurveda.

Aegle marmelos

Aesculus hippocastanum Linn.

Family : Hippocastanaceae
Trade/Common Name : Horse-chestnut

A perennial tree, deciduous, long stalked palmate leaves and Flowers are arranged in narrow-pyramidal clustered of white (red-streaked) in clour and blooming in the month of April to continue upto May.

It is an avenue tree and indigenous to the hills of Balkan Peninsula. It occurs in roadsides, woods, exposed areas in South-West Asia and the Himalayas. It is also cultivated in various areas of low-land of mountains of Kashmir. It is propagated by seeds.

It contains saponin upto 10%, coumarin aesculin, aglycone and tanin. The fruits/seeds used as a cataplasm for mammary indurations and other types of cancer. Barks/nut infusions used for healing ulcers. The astringent bark is said to be sternutatory, tonic, narcotic and febrifuge hence useful in malaria. Fruits have been used for backache, neuralgia, rheumatism and whooping cough; also for haemorrhoids and other rectal complaints. Flowers are said to be anodyne, astringent, tonic.

Aesculus indica Colebr.

Family : Hippocastanaceae
Trade/Common Name : Indian horse-chestnut

A deciduous tree, to 40 m high with scaly buds. Leaves opposite, digitate, 5-9 foliate, estipulate; petioles 10-15 cm long; leaflets oblong or oblanceolate, serrulate, acuminate. Flowers pinkish-white, in terminal, erect, long panicles and blooming in the month of April to continue up to June.

It is found in moist Oak-Rhododendron forests of tropical and sub-tropical Himalayan region between 1800-3000 m altitude. It is propagated by seeds.

The seeds contains a mixture of saponins, as aescine and also contains flavonoid, glycosides, aesculine, albumin and fatty oils. The seeds are used for the complaints of the veins, such as phlebitis, haemorrhoids, varicose veins; in ulcers; to prevent thrombosis.

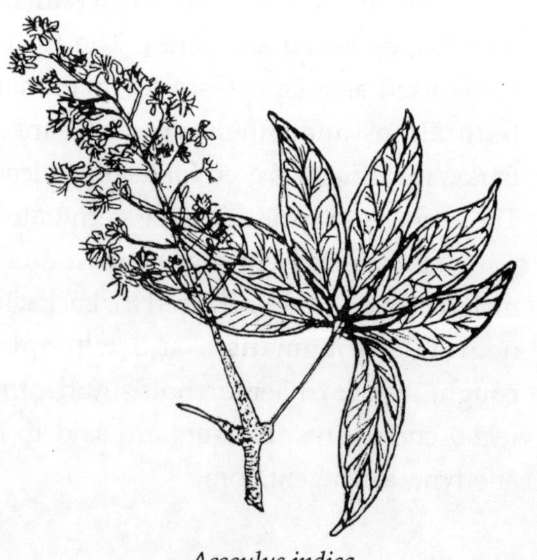

Aesculus indica

Agave sisalana Perr.

Family : Agavaceae
Trade/Common Name : Sisal

A perennial succulent herb upto 1 m tall. Leaves dense rosette, stiff and spreading in all directions, flat or slightly concave, sword shaped thick, margins smooth or bear a few small teeth. Epidermal part leathery, grey green with conical terminal short spines. Infloresence emerges from center of the rosette, tall. Flowers are long, tulip-shaped, green, ill smelling and appear in clusters at the forked tips of the 30-40 horizontal branches.

Agrimonia eupatoria Linn.

Family : Rosaceae
Trade/Common Name : Agrimony

A perennial herb upto 60 cm in height. Leaves are bepinnate with denatate margins, hairy all over, more towards midribs. Stem is unbranched with a long flowering spike bearing yellow flowers.

It is found in pastures, woodland clearings, roadsides and waysides in forest of temperate zones of North Hemisphere. Occasional in western Himalaya (Kashmir & Kumaon region of Uttaranchal). It is propagated by seeds.

Agrimonia eupatoria

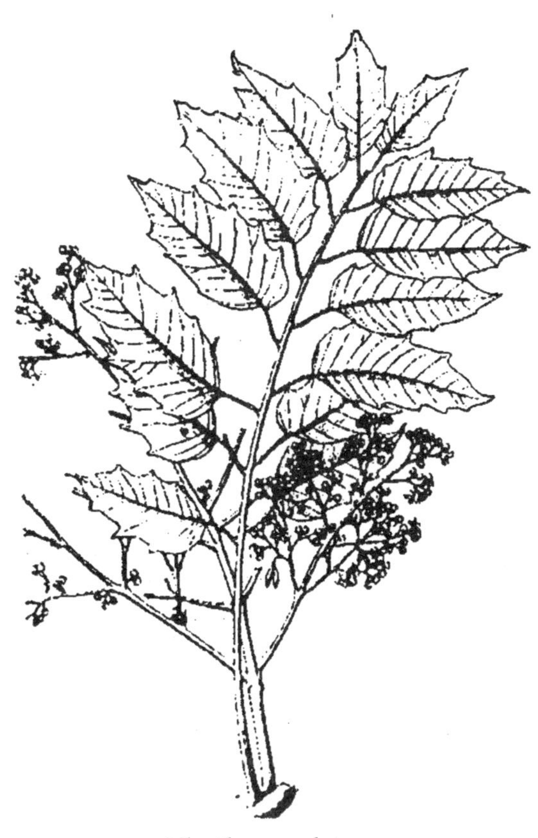

Ailanthus excelsa

It contains tannin, flavone dyes and traces of essential oils. The drug has fairly good aromatic odour and bitter taste. In France, agrimony tea is considered to be astringent, antiinflammatory, regulating stomach and liver funtions. Locally, the herb is dried and the water extract is given to relieve gastro-intestinal disturbances.

Ailanthus excelsa Roxb.

Syn. *Pongelion wightii* Tiegh.

Family : Simaroubaceae

Trade/Common name : Tree of heaven.

A lofty deciduous tree, to 25 m high, with pale-white stem. Leaves pinnate, 30-100 cm long, glandular hairy; leaflets 10-15 pairs, 10-15 cm long, alternate or subopposite, coarsely and irregularly serrate, ovate or falcate, lanceolate, oblique at base. Flowers yellowish in large lax, much branched, panicles and blooming in the month of April-June.

It occurs wildly in some forest of Deccan, Bihar, Jharkhand, Western Ghats from N. Kanara and Mysore to Travancore; cultivated in road-side and garden. It is propagated by seeds.

Bark contains triacotane, hesatriacotane and a non-glycosidal bitter, excelsin. It yields beta-sitosterol, vitexin and malanthine. Leaves are specially useful in asthma, bronchitis, dyspepsia in the treatment of weakness after child birth. Its paste is applied as poultice in erysipelas, goiter, swelling and ulcers. Bark is aromatic, bitter, tonic, febrifuge, expectorant and antispasmodic; used for dyspeptic complaints and as astringent in diarrhoea and dysentery and given in chronic bronchitis and asthma.

Ajuga bracteosa Wall. ex Benth.

Family : Lamiaceae
Trade/Common Name : Neelkanthi

Ajuga bracteosa

A perennial, erect or ascending hairy herb, often prostrate, with oblanceolate or sub-spathulate, obtuse, sinuate toothed leaves; purplish violet tinged from lower surface. Flowers are purplish-violate, tinged from lower surface in distant, axilliary whorls in a spike and blooming in the month of April-August.

Globally, It is distributed from Afghanistan to Bhutan, India, China, Japan, South East Asia. It is found as a weed in field borders, along roadsides, open slopes and even on rock crevice upto 1500 m altitude. It is propagated by seeds.

The leaves contain glycosides and tannin and also yield ceryl alcohol β-sitosterol, β-sitosterol and palmitic acid. The leaves are used as stimulant, diuretic, tonic, apparent and febrifuge. Locally, water extract of fresh leaves is generally administered orally with sugar to children. The decoction of dried leaves is used against stomach pain.

Albizia lebbek Benth.

Family : Mimosaceae
Trade/Common Name : Indian Walnut

A large, deciduous tree, bark brownish gray, rough with numerous, small cracks; leaves bipinnate; leaflet 4-9 pairs, unequal sided, dark green, slightly pubescent beneath. Flowers are white, fragrant, in globular, umbellate heads on short pedicels and blooming in the month of April-May.

Albizia lebbek

It is found throughout India, much planted as avenue tree and also in tea gardens for shade. It occurs scattered in mixed deciduous forests and ascending to 1200 m in the Himalaya. It is propagated by seeds.

The bark, leaves, flower and seeds are medicinally important. They contain gum, saponin and tannin and used as alternative, aphrodisiac, astringent, expectorant, restorative and tonic.

Allium cepa Linn.

Family : Liliaceae
Trade/Common Name : Common onion

A bulbous herbaceous plant upto 1.2 m in height. Outer coat is membranous. Leaves cylindrical except for groove on inner surface, hallow, upto 15 cm long. Umbels are large, pedicle with bracteoles and many times longer than flowers.

It is grown throughout the various parts of India. It is propagated by seeds.

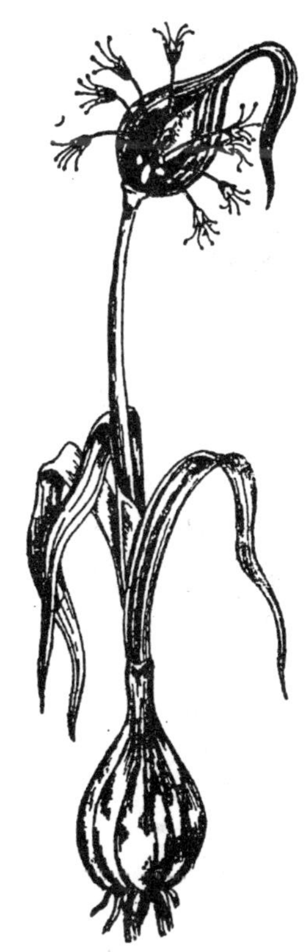

Allium cepa

Bulbs yield volatile oil (0.018-0.04%) through steam distillation. It is pungent with lasting colour. The oil contains disulfide. Its fresh juice contains thiocyanic acid and thiocyanate. The oil is used in culinary preparation for flavouring soups, table sauces, meats etc.

Allium sativum Linn.

Family : Alliaceae
Trade/Common Name : Garlic

An annual or perennial, bulbous plant with numerous white bulbs called as cloves enclosed in a common, membranous white or pinkish covering/ sheath. It has long, flat,

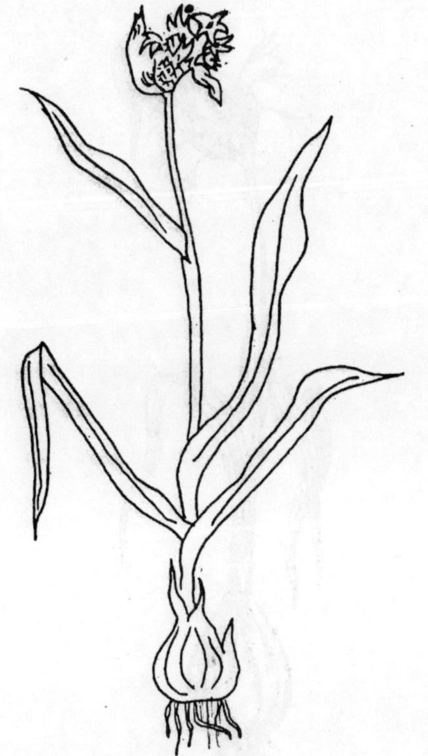

Allium sativum

narrow leaves and bears small, white flower and bulbils in a globose mass on a solid scape and blooming in the month of May-June.

It is cultivated as a kitchen garden crop in most part of the country. It is propagated by seeds.

The bulb contains allin, allicin, alitin, seordinine, scordine, anthocyanins and essentioal oil. The garlic has proved to be an antibiotic, antiseptic, antidiabetic, antiarthritic, anticholesteremic, anti cancer, antibacterial, antifungal, insecticidal and pesticidal.

Aloe barbadensis Mill.

Syn. *A. vera* Tourn. ex Linn.

Family : Lillaceae

Trade/Common Name : Indian Aloe

A perennial fleshy plant. Leaves sessile, crowded, lanceolate, erect, spreading, spiny toothed at the margin, fleshy. Flower; on a scape, scaly branched and longer than leaves, yellow in colour and blooming in the month of July-September.

It is found in hotter provinces in India. Many of the form of this species are naturalized in India and semi-arid regions and dry westward valleys of the Himalaya. It is propagated by suckers.

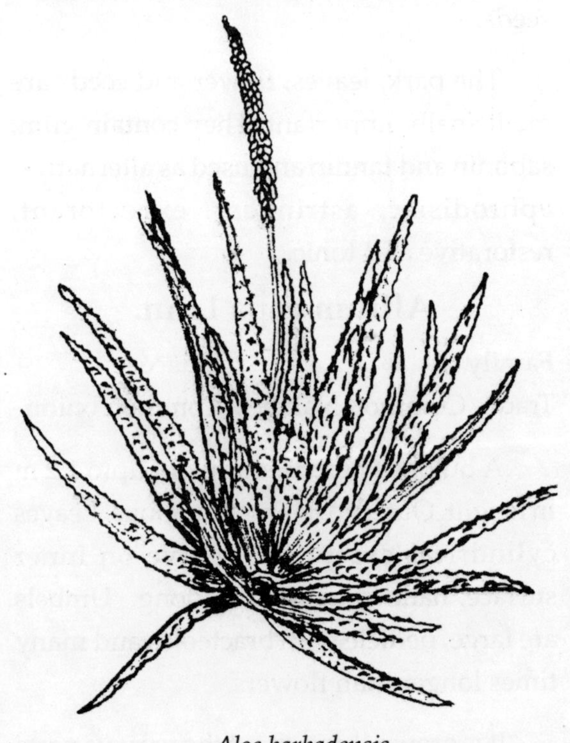

Aloe barbadensis

The pulp contains glucoside, barbaloin, b-barbaloin and aloe-emodin. And used as bitter, purgative, stomachic, alternative, aphrodisiac, anthelmintic, emmenagogue, cathartic and blood purifier.

Ammi visnaga Linn.

Syn. *A. majus* Linn.

Family : Apiaceae
Trade/Common Name : Bishop's weed

An annual herb upto 120 cm with long taproot and a solid, erect stem. Leaves decompound, light green, alternate, pinnately divided, lanceolate. Flower white, borne on auxillary and terminal compound umbel.

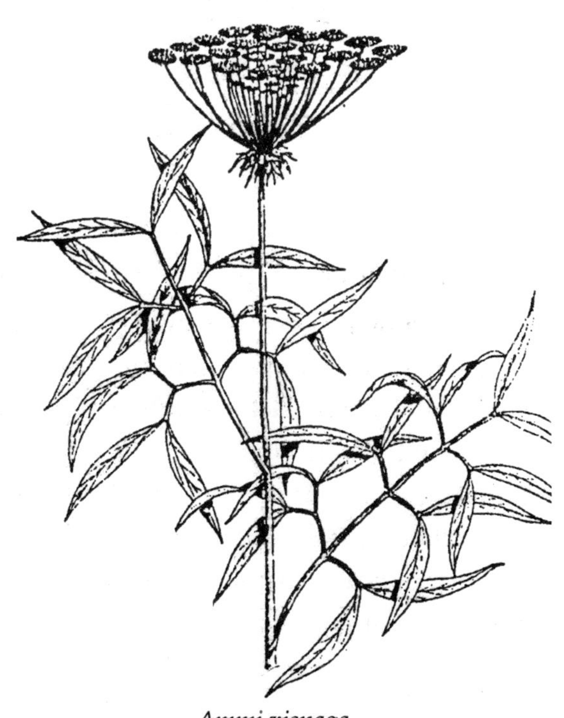

Ammi visnaga

It is native of Europe. It is found in Kashmir, Uttaranchal, Punjab and now cultivated in hot climate of North India (Madhya Pradesh, Chattisgarh, Rajasthan etc). It is propagated by seeds.

The seeds contain khellin (2%), visnagin (0.1%), xanthotoxin (0.4%) and khellol glucoside (0.3%). The acive principle khellin is a selective coronary vasodilator and bronchodilator, used in the treatment of coronary insufficiency, angina pectoris and brochial asthma. Locally, decoction of fruits is used to treat urethral spasm to cause passage of kidney stones and also for renal colic.

Anagallis arvensis Linn.

Family : Primulaceae
Trade/Common Name : Pimpernel

A annual herb with branches arising from the base, gland dotted, upto 30 cm long. Leaves simple, rarely whorls. Flowers are mostly reddish, sometimes blue (rarely) and blooming in the month of June-July.

It is found as a weed in grasslands, lawns, orchards, wheat field, paddy field borders in North-West India and Centeral India. It also introduced into temperate regions of Himalayas. It is propagated by seeds.

It contains saponin, cyclamins etc. and used as expectorant, stimulant, diaphoretic. In indegenous medicine used for the treatment of gout, dropsy, cerebral

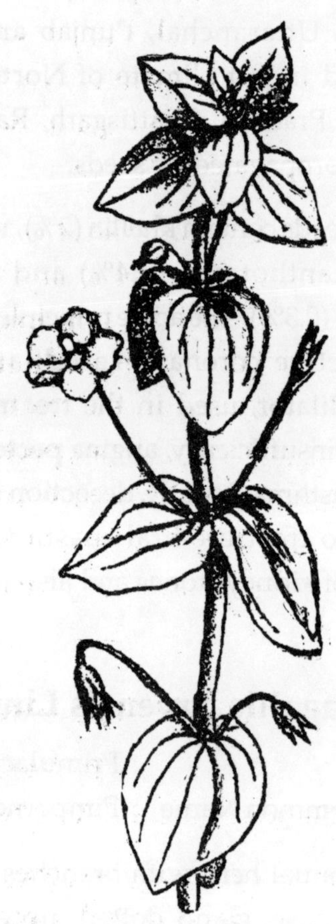

Anagallis arvensis

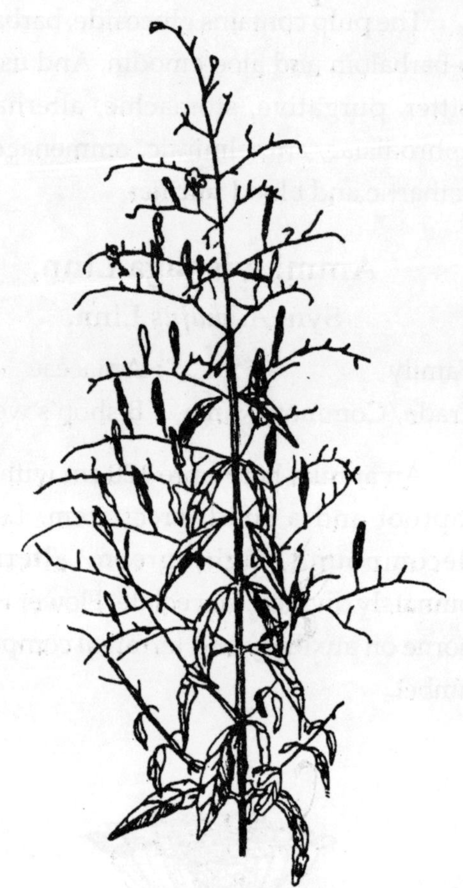

Andrographis paniculata

affections, leprosy, hydrophobia and also used to detach leeches from the body.

Andrographis paniculata Wall. ex Nees.

Family : Acanthaceae
Trade/Common Name : Kalmegh

A much branched, erect, annual herb. Branches 4-angled. Leaves lance-shaped. Flowers small, rose coloured and arises in large spreading sparse branches.

It is native to tropical Asia. It occurs throughout India. It is propagated by seeds.

It is known as 'King of bitters'. It is the source of several diterpenoids of which the bitter water soluble lactone andrographolide is important. It possess bitter, tonic, febrifuge, antidysentric, antiperiodic properties. It is effective against chronic malaria, jaundice, anemia, bowel complaints and loss of appetite.

Anethum graveolens Linn.

Family : Apiaceae
Trade/Common Name : Dill

A annual, erect, glabrous herbs, 50-150 cm tall. Leaves 3-4 pinnate; lower leaves petioled, upper ones sessile, smaller, 2-3 pinnate; segments filiform; sheath white margined, umbels terminal or leaf opposed; peduncle 5-15 cm long. Involucre and involucels. Umbellules 8-10, many flowered; pedicels long, yellow in colour and blooming in the month of February to continue up to June.

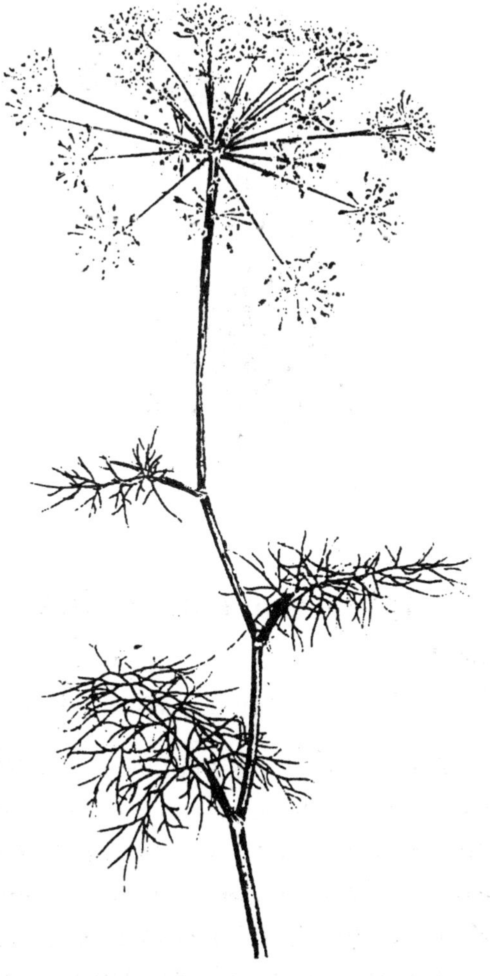

Anethum graveolens

It is grown in plain Gangatic areas and Central India. It is propagated by seeds.

Whole plant but mainly seeds (Fruit) yield an aromatic oil known as Dill oil containing dillapole (15.6-39.6%) and carvone. Its leaves are used for seasonal soups, sauces and particularly pickles. The seeds used as carminative, aromatic, stimulant, and diuretic in Ayurvedic and Unani medicines. The emulsion of seed oil in water is an important constitutent of grape water, useful in flatulence, colic pain, vomiting and hiccups

Angelica glauca Edgew.

Family : Apiaceae
Trade/Common Name : Chora

A tall, robust, erect, aromatic, perennial herb 1.0-1.5 m tall with thick, stout, perennial rootstock. Roots are brownish, long, fleshy branching, thick, annulate, fusiform, resinous and pungently aromatic. Leaves are pinnately divided the basal ones are ternate pinnate. Leaflets oval to irregularly mucronate, serrate. Flowers white or purple in long stalks compound umbels and blooming in the month of July-August.

It is found in the shady and moist slopes of western Himalaya from Kashmir to Uttaranchal at the altitude 2700-3400 m. It is propagated by seeds.

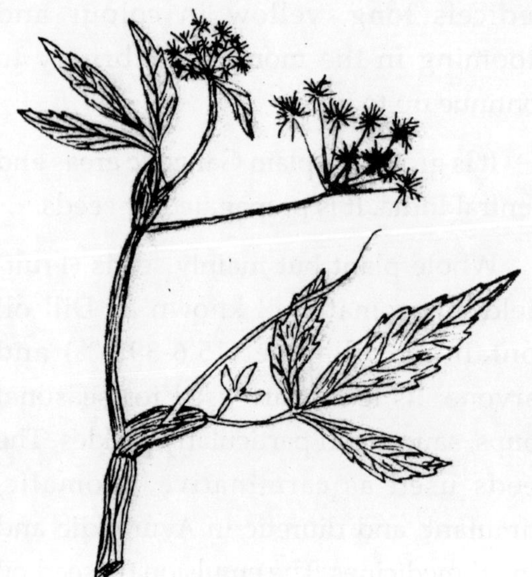

Angelica glauca

The dry roots yield an essential oil (1.3%), resin and valerianic acid and several furocoumarins such as angelicin, bergapten, xanthotoxin, umbelliferene. The roots are used as pungentive, aromatic, stomachic, stimulant, carminative, diaphoretic, appetitive and diuretic.

Antirrhinum majus Linn.

Family : Scrophulariaceae

Trade/Common Name : Snapdragon

A erect, annual, densely sticky-pubescent herbaceous plant upto 0.9 m in height. Leaves ovate, glossy and mid-green in colour. Flowers borne in terminal racemes, purplish red or white in colour.

It is native of Mediterranean region. It is propagated by seeds.

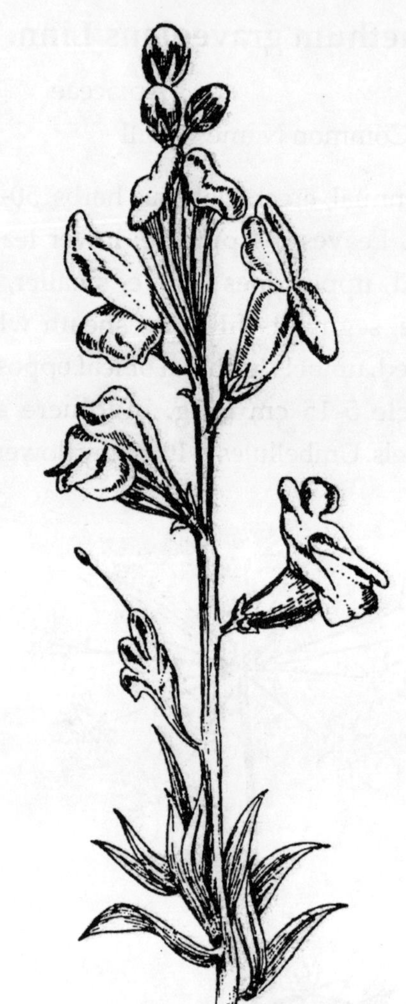

Antirrhinum majus

Apium graveolens Linn.

Family : Apiaceae

Trade/Common Name : Wild Celery

An annual or biennial, erect, glabrous herbs with fistular, angular stem. Leaves pinnate, leaflets 1-6 cm long, oval-obovate, usually 3-lobed incised, glabrous; sheath white margined; petioles to 1 cm long or lacking in upper leaves. Umbels leaf opposed, 12-15 rays. Involucre and involucel

Apium graveolens

absent. Umbellules 20-25 flowered, white in colour and blooming in the month of March to continue up to June.

In India, it is cultivated in hilly areas. It is propagated by seeds.

Seeds yield an essential oil known as Ajmod or celery oil. It is used in the flavouring of all kinds of prepared foods such as soups, meats, pickles and vegetable juices. It is also used in perfumery and pharmaceutical industries. The oil is also used in compounding Ayurvedic formulations. While, fatty oil of seed is antispasmodic and nerve stimulant, seeds of celery are rich in vitamin–B.

Archangelica officinalis Hoffm.

Syn. *Angelica archangelica* Linn.

Family : Apiaceae
Trade/Common Name : Angelica

A perennial herb upto 2 m tall. Rootstock thick and very much branched.

Archangelica officinalis

Flowers white in large gigantic compound umbels and blooming in the month of July to continue upto August.

It occurs along slow moving streams, damp meadows, forest gullies in rocky substare of Western Himalaya in 2500-3800 m elevation. Globally, it is distributed in temperate Europe and Siberia; Grassland and Lceland to Kamchatka and the Aleutians.

Arctium lappa Linn.

Family : Asteraceae
Trade/Common Name : Jangli Kuth

An erect, branched, biennial to perennial upto 1 m or more tall. Leaves

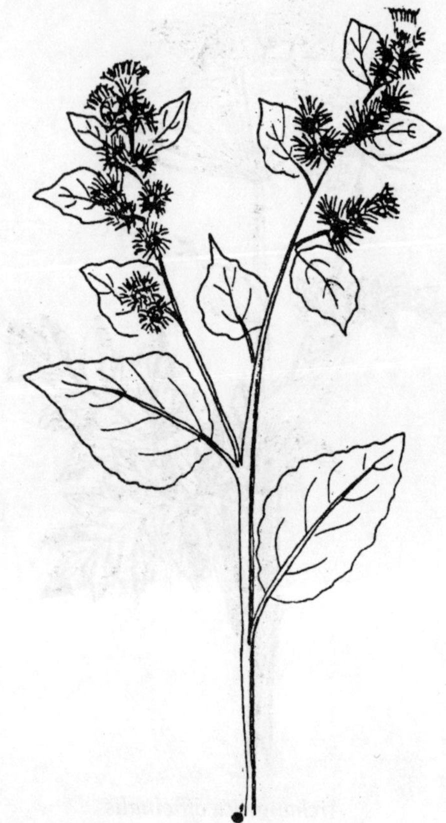

Arctium lappa

ovate, cordate, margins undulate, usually cottony beneath, stalked. Flower heads globose in terminal clusters and blooming in the month of June-August.

It is found widely in the Western Himalaya at altitude of 1800-3800 m. It is propagated by seeds.

The roots contain acetic acid, butyric acid, caffeic acid, chlorogenic acid, lauric acid, etc. It has diuretic and arexigenic properties and it has been used for cutaneous eruptions, rheumatism, cystitis, gout and specifically for eczema and psoriasis.

Arnebia benthami (Wall. ex G.Don.) John.

Syn. *Macrotomia benthamii* (Wallich) A. DC.

Family : Boraginaceae
Trade/Common Name : Ratanjot

An erect, robust, herbaceous perennial 30-90 cm. Tall with stout, woody, deep-red roots. Leaves narrow, lanceolate, hairy. Flowers pink purple in dense terminal spikes and blooming in the month of July-August.

It is found from Pakistan to West-Nepal between 3300-4500 m in open slopes and

Arnebia benthami

alpine pastures of Himalayan region. It is propagated by seeds and by division of roots with stem portion in tact.

The herb contains essential oil and a red dye. The plant is used as expectorant, antiseptic, antibiotic and used for cardiac disorders.

Arnebia euchroma (Royle ex. Benth.) John.

Family : Boraginaceae
Trade/Common Name : Ratanjot

An erect, caespitose, perennial, hairy herb, with thick purple roots. Stem many, arising from the axil of the leaves, which formed a sterile basal cluster the preceding year. Basal leaves lanceolate, acute; lower and middle cauline leaves sessile, lanceolate; upper ones usually shorter and broader. Flowers pink or purplish-white, in terminal, subglobular inflorescence and blooming in the month of July-August.

It occurs from Afghanistan to Central Nepal, Western and Central Asia between 3300-4500 m in drier areas. It is propagated by seeds.

The roots contain Ac-Shikonin and 4 related compounds. It has been reported to be an antipyretic, anti-microbial and anti-inflammatory and used in toothache, ear-ache, eye diseases, as healer of cuts and wounds and also the fire burns.

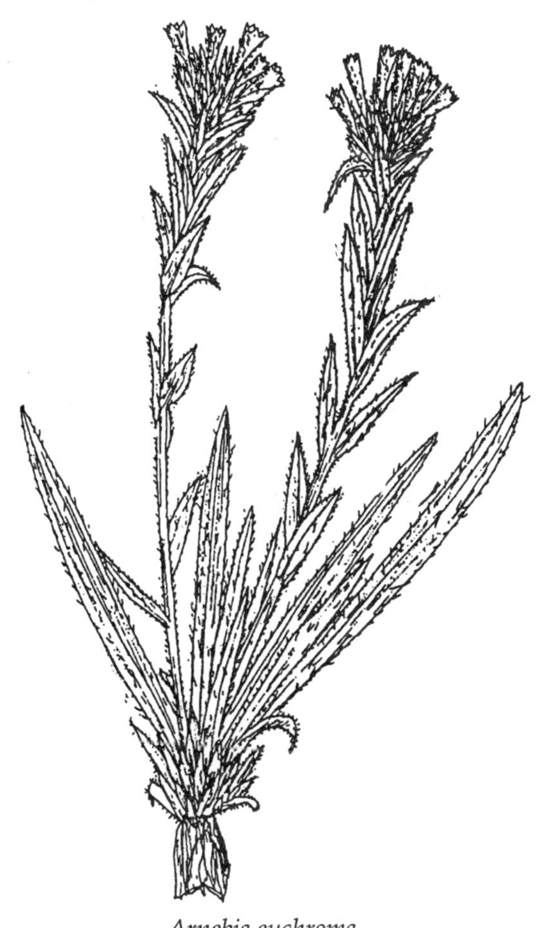

Arnebia euchroma

Artemisia absinthium Linn.

Family : Asteraceae
Trade/Common Name : Madder Wood

A aromatic, biennial herb, branched with dissected leaves. Stem upto 60 cm tall. Flower heads yellowish red arranged in clusters of axillary spikes and blooming in the month of July continue upto August.

Globally, it distributed in North Asia, Afganishtan and westwards to Atlantic. In

Artemisia absinthium

India, it occurs as weed in cultivated fields in temperate regions. It is propagated by seeds.

It contains absinthin, bitter glycoside, essential oil. Main source of durg 'Afsantheen' used in India in chronic fevers, swellings and inflammation of liver. Oil is a fragrant component used in creams and detergents. Locally, the aerial herbs is cut and dried to be used as a home remedy against round worm.

Artemisia annua Linn.

Family : Asteraceae
Trade/Common Name : Worm wood

A annual, erect herb upto 2 m tall with slender, glabrous grooved branches. Leaves pinnatisect with serrate segments. Flower heads subglobose, arranged in slender panicled racemes and blooming in the month of July to continue upto August.

Globally, it distributed in China, South-Eastern Europe, Bulgaria, Romania, Yugoslavia, Uzbekistan, Russia and Centeral Asia. Introduced to India. Cultivated and naturalized in Kashmir. It is propagated by seeds.

Leaves contain artemisin (0.06-0.7%). Flowering heads contain essential oil

Artemisia annua

(2.5-3.5%). It is used in perfumery, cosmetics and dermatology and also having fungicidal properties.

Artemisia maritima Linn.

Syn. *A. brevifolia* Wall.

Family : Asteraceae

Trade/Common Name : Worm weed, Santonin Plant

A strongly aromatic, much branched shrubby perennial with highly dissected, grayish to white leaves. Flower heads yellow to yellowish-red arranged in clusters of axilliary spikes and blooming in the month of July-August.

It occurs commonly in rocky substrate, some times in saline tracts in dry temperate and cold desert areas of Western Himalayan regions (Kashmir to Uttaranchal). It is propagated by seeds.

The flower heads contain santonin (1-1.6%) and essential oil having cineol (9.3%). It is used as aromatic, bitter for the loss of appetite, week digestion and stomach complaints. Locally, the leaf extract in water is given in very small doses against intermittent and remittent fevers. The water extract is used against skin eruptions as external use.

Artemisia pallens Wall.

Family : Asteraceae

Trade/Common Name : Davana

An erect, hoary, herb with broadly ovate decompoundly, pinnalisect leaves, heads

Artemisia maritima

Artemisia pallens

broadly hemispheric nodding, subsecund, distant in slender panicled racemes.

It is grown in mainly South India. It is propagated by seeds or cuttings.

Leaves and flowers yield an essential oil known as Davana oil. It is used in perfumery, cosmetic and food industries. It is also uses in flavouring of tobacco and some costly beverages.

Asparagus adscendens Roxb.

Family : Liliaceae
Trade/Common Name : Safed Musli

A struggling or scandent, much branched spinous shrub with woody terete stems. Cladodes long, linear, spines, stout, straight, densely, crowded; flowers small, white mildly fragrant in solitary or fascicled, simple or branched, copious racemes and blooming in the month of October to continue upto February.

It occurs throughout tropical and sub-tropical parts of India ascending upto 1300 m in the Himalaya. It is propagated by seeds.

The roots contain asparagines, saponin (steroidal). Fruits contain b-sitosterol, sarsasapogenin and diosgenin. It is used as appetizing, diuretic, aphrodisiac, laxative, astringent useful in dysentery, diarrhoea, throat complaints and leprosy.

Asparagus racemosus Willd.

Family : Liliaceae
Trade/Common Name : Satawar

It is straggling or scandent, much branched perennial, spinose, undershrub.

Asparagus adscendens

Asparagus racemosus

Root stock fusiform, tuberous, fasiculated. Stems triquetrous with patent or recurred spines. Leaves scaly, triangular, stift acuminate, reduced to suberect or subcurved spines. Flowers white in axillary, solitary or fascicled, simple or branched racemes terminated by flowers born in a compact bunch, fleshy & spindle shaped. Its marked in pieces 5-15 cm in length & 2 cm in thickness. Silvery white or light ash-coloured externally and white internally. Lack a well-marked odour, but are sweet & bitter in taste.

It is found in Sal and mixed forest of Madhya Pradesh, Chhatisgarh and Jharkhand. It is propagated by seeds or slips.

Tuberous roots contains four saponins, *viz.* Shatavarin I to IV. Shatavarin is a glycoside of sarsasapogenin. Flower contains quercetin, rutin (2.5% dry basis) & hyperoside. Dried tuberous roots has ulcer healing resistance or cytoprotection. It has also been identified as one drug to control the symptoms of AIDS. Alcoholic extract of the roots increased the weight of mammary glands in post-partum and estrogen-primedrats. It is said to be tonic and diuretic and used as galactogogue & aphrodisiac.

Astilbe rivularis Buch.-Ham. ex D.Don.

Family : Saxifragaceae
Trade/Common Name : Buriokhati

A perennial, patent shrub or undershrub with thick rootstocks up to 1.5 m high. Leaves alternate, 15-35 cm long, 2-pinnate, lower ones with 3-leaflets; leaflets ovate, acuminate, base rounded or cordate, sharply and irregularly toothed, glabrescent above, hoary pubescent. Flowers minute, greenish-yellow in colour, arranged in lax terminal panicle and blooming in the month of September to continue up to November.

It occurs shrubberies, sunny moist slopes of temperate regions in Sikkim Himalaya in between 1600-3300 m altitute range and rarely found inside the forest. It is propagated by seeds.

The root's powdered is simply taken orally during diarrhoea and dysentery.

Atropa acuminata Royle ex Lindl.

Syn. A. belladonna Linn.

Family : Solanaceae
Trade/Common Name : Deadly night shade

A tall, straight plant about 1 m high. Leaves are stalked, elliptic, lanceolate, acuminate 7-16 cm long and 5-10 cm broad. The aerial parts die every year in autumn and the new ones appear in the following year. Flowers are yellowish coloured, solitary shortly stalked and blooming in the month of June-August.

It is found in open places from Pakistan to Himachal Pradesh at altitude 2500-3000 m. It is propagated by seeds and root cutting.

The roots contain atropine, hyocyamine, hyoscine, cuscohygrine and fatty oil (25%).

Atropa acuminata

It is used for the manufacture of tinctures, plasters etc. and used in asthma, whooping-cough bladder and urethral spasms. It is considered as sedative and narcotic.

Azadirachta indica A. Juss.

Family : Meliaceae
Trade/Common Name : Neem Tree

A tall tree with pinnate leaves i.e. its leaves are divided into numerous smaller segments called leaflets, each looking like an ordinary leaf. Flower are small, white, in short axilliary bunches and blooming in the month of March-April.

It is found throughout India. As a common road-side plant, grown over the greater parts of India. It is propagated by seeds.

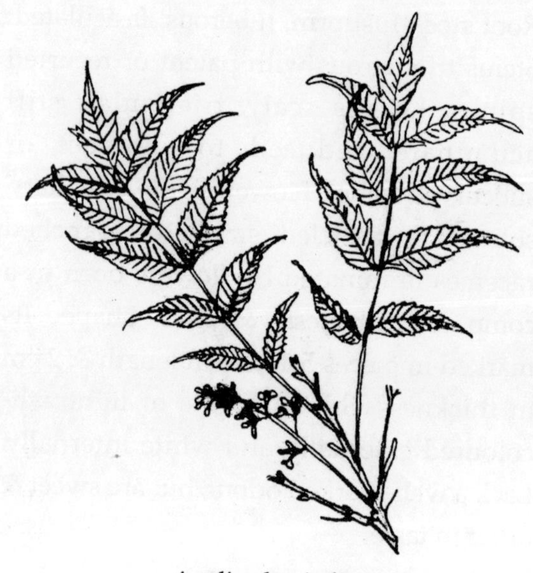

Azadirachta indica

The whole plant viz. bark, root bark, young fruits, nut or seed, flowers, leaves, gum, toddy, etc. is medicinally important and used in leprosy, intestinal worms, piles and urinary diseases. Root bark and young fruits are astringent, tonic and antiperiodic. Bark is bitter, tonic, astringent and vermifuge. Fruit is purgative, emollient and anthelmintic. Leaves are discutient, leaf juice anthelmintic.

Bacopa monnieri (L.) Penn.

Family : Scrophulariaceae
Trade/Common Name : Jal Brahmi

A prostarte, succulent herb with branches spreading or ascending and rooting at the nodes. Leaves obovate-oblong or spathulate, upto 18 mm long. Flowers campanulate, bluish-purple or white with bluish veins, solitary, axillary, short-or-long-

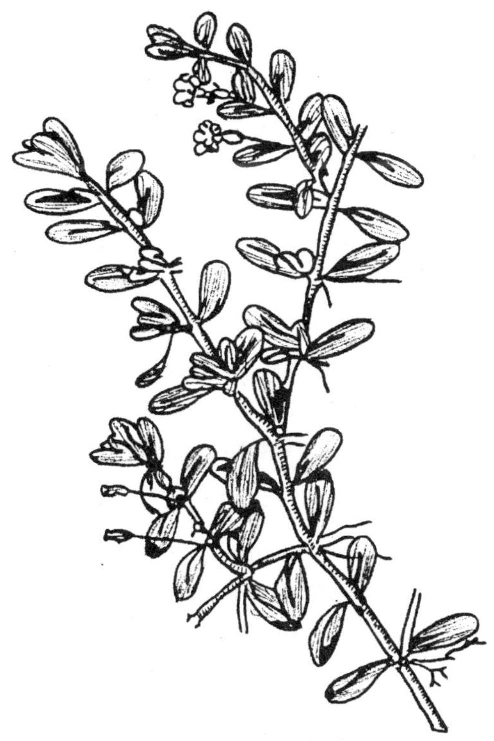

Bacopa monnieri

pedicilate and blooming in the month of Agust to continue upto October.

It is found in marshy tracts in Sub-tropical region, upto 1000 m elevation. It also found in West Bengal. It is propagated by runner.

The whole herb is used in Ayurvedic, Unani and Siddha systems of medicine as a nerve tonic, diuretic and blood purifier. It is used as "Memory Plus".

Baliospermum montanum Muell-Arg.

Family : Euphorbiaceae
Trade/Common Name : Danti

A leafy, stout usually monoecious under shrub 0.9-1.8 m high, with herbaceous branches arising from the root. Leaves are sinuate toothed; upper small, lanceolate, penninerved, and lower large, reaching 15-30 cm long, often as broad, ovate. Flowers numerous in axilliary racemes or contracted panicles and blooming almost all the year around.

Baliospermum montanum

It is found as naturally in Bihar, Chattisgarh, North Bengal, Nagpur and outer Himalaya from Kashmir to Bhutan. It is propagated by seeds.

The dries root contains b- sitosterol and seeds are rich in oil, resin, starch, etc. It is pungent, heating, purgative, anthelminthic, diuretic, alexiteric and useful in pains, diseases of the skin and of the abdomen, in piles, enlarged spleen, itching, anemia, leucoderma and jaundice. It is digestive and appetizer and is best for ascites, hernia, oedema and other diseases due to vitiation of Vata and Kapha.

Bauhinia vahlii W. & A.

Family : Caesalpiniaceae
Trade/Common Name : Mahul

A large extensive woody climber with many branches. Leaves are coarse and thick upto 25 cm long. Flowers are white-yellowish, appear in many flowered terminal clusters.

It is found in different parts of India and also at Karean in the Chhindwara district of M.P. It is propagated by seeds or cuttings.

The root decoction about 50 ml is taken three times a day for four days in fever.

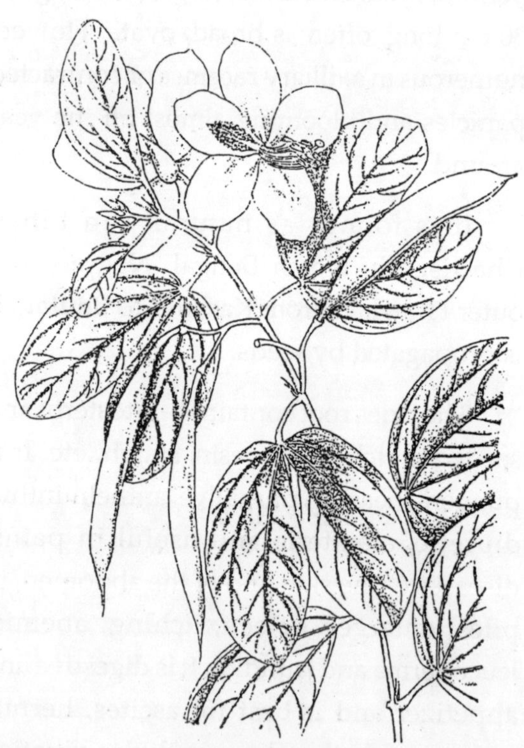

Bauhinia vahlii

Bauhinia variegata Linn.

Family : Fabaceae
Trade/Common Name : Kachnar

A fairly medium sizes to a large tree. Leaves deeply cordate, 11-15 nerved, sub-coriaceous, cleft through about a quarter, lobes obtuse. Flowers are large, fragrant, sessile or shortly pedicellated in few flowered lateral corymbs and blooming in the month of April when leafless.

Its occurs in semi-decidous forests and roadsides in Sub-Himalayan tract from the Indus east-wards, also dry forests to Eastern, Central and South India. It is propagated by seeds.

The root, bark, flowering buds and flower are medicinally important. The flower contains hentriacontene, octacosanol, reducing sugars, stigmasterol, β-sitosterol,

Bauhinia variegata

Abelmoschus moschatus

Abutilon indicum

Acacia catechu

Acacia nilotica

Achyranthes aspera

Aconitum heterophyllum

Aconitum violaceum

Acorus calamus

Aegle marmelos

Ailanthus excelsa

Ajuga bracteosa

Albizzia lebbeck

Allium cepa

Allium sativum

Aloe barbadensis

Andrographis paniculata

Asparagus racemosus

Azadirachta indica

Bacopa monnieri

Bauhinea variegata

Berberis lycium

Bombax ceiba

Butea monosperma

Calotropis gigantean

Cannabis sativa

Carthamus tinctorius

Cassia fistula

Catheranthus roseus

Centella asiatica

Chenopodium ambrosioides

etc. The bark is anti-inflammatory, useful in skin diseases, ulcers and scrofula. Dried buds are used in dysentery, piles and worms. Roots are used as antidote to snake poisoning. Decoction of roots is used in dyspepsia.

Berberis aristata DC.

Family : Berberidaceae
Trade/Common Name : Rasont

A large, thorny, deciduous shrub nearly 2-5 m tall with pale rough, furrowed bark; leaves stalked, ovate or elliptic, fascicled in the axils of simple or branched spines, leathery, often with sharply toothed margin; flowers yellow, in compound racemes with stout pedicles and blooming in the month of May-June.

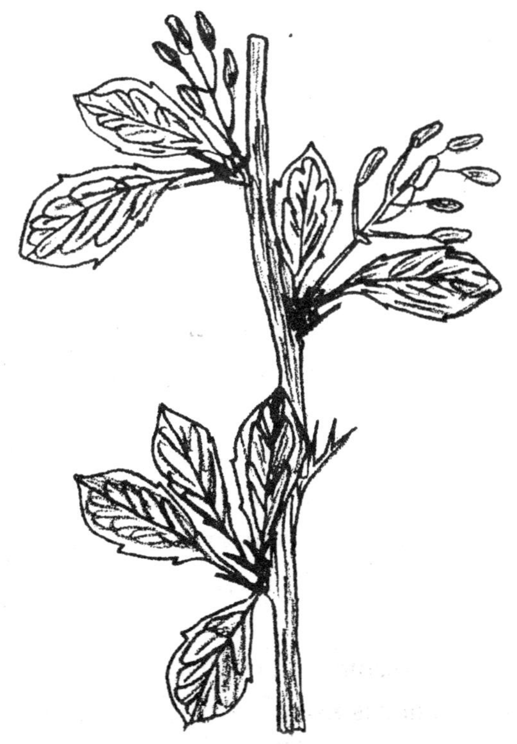

Berberis aristata

The plant is found throughout the country between 1600-3300 m altitude but rare and sporadic conditions in shady ravines, field borders places. It is propagated by seeds.

Berberine is the principle alkaloid and also contains oxycanthine, umbellatine, fat, tannin, and resin.The wood, rootbark and extract have been used in Hindu medicines from a very remote periods and used as alternative, deobstruent and used in skin diseases, menorrhagia, diarrhoea, jaundice and eye.

Berberis lycium Royle

Family : Berberidaceae
Trade/Common Name : Daruhaldi

It is a large, spiny, evergrren shrub of average height. Twigs are pale yellowish, bark grayish, rough, rather deeply furrowed: leaves are lanceolate or nearly obovate, oblong, entire. Flowers, yellow in a simple raceme and blooming in the month of March-April.

It occurs in sub-tropical and temperate regions from Kashmir to Uttaranchal on the outer Northern-Western Himalayas on clearances and along roadsides between 800-2500 m in altitude. It is propagated by seeds.

Berberine and Umbellitine are the main alkaloids and fruits contain malic, tartaric, citric acids and tannin. The roots is

Berberis lycium

considered bitter with unpleasant taste and used in splenic troubles, tonic, febrifuge, intestinal astringent: good for cough,chest and throat troubles, eye sores and itching of the eye: piles, menorrhagia: useful in chronic diarrhorea.

Berberis vulgaris Linn.

Family : Berberidaceae

Trade/Common Name : European Berberry

A perennial shrub grows upto a height of 3 m. Colour of inner wood of the stem is yellow. Leaves are small, upper surface green and lower surface grey, leaves on young sprouts are spiny. From each nodes sharp spines appear. Flowers are yellow in colour.

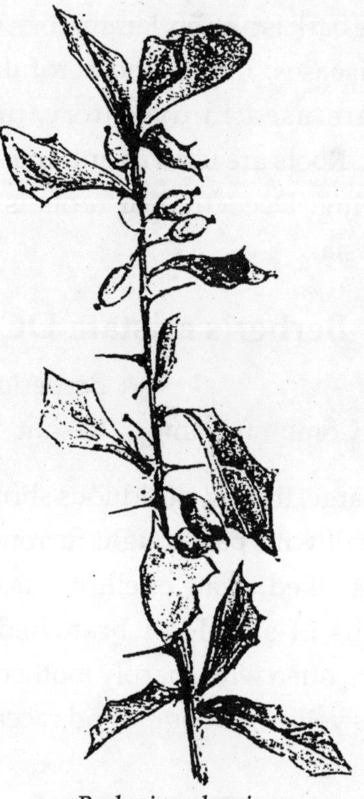

Berberis vulgaris

It is native to Britain. It is distributed in various parts of India. It is propagated by seeds or cuttings.

Stem and root bark have medicinal properties. The fruit is rich in vitamin C. A wine is prepared from berberry fruits and used in malignant, choleraic and pestilential fever, diarrhoea and menstrual fluxes. Squeezed blooms are applied to clean old ulcers. The concentration juice of berries is used for gums and teeth trouble. The root and wood which are bitter in taste is used as purgative tonics. The bark has been used to treat dysentery and indigestion. The plant preparation is also used for jaundice, dropsy and congestion of liver and spleen.

Bergenia ciliata (How.) Sternb.

Syn. *B. ligulata* (Wall.) Engl.

Syn. *Saxifraga ligulata* Wall

Family : Saxifragaceae
Trade/Common Name : Pashanbheda

A rhizomatic, perennial herb. Rootstock thick, dark brown from outerside, pinkish from within: covered with dark sheaths of withered leaves. Stem short, thick and fleshy; leaves stalked and sheathing at the base; ovate or orbicular cordate; margin fringed with short and stiff hairs. Flowers are white, pink or purplish in terminal corymbs and blooming in the month of March-April.

The plant is found throughout temperate Himalayas from Kashmir to Bhutan 2700-3000 m in altitude on moist, rocky slopes and more frequently on rocky boulders in ravines and nalahs. It is propagated by rhizome cuttings.

The rhizome contains gallic acid, tannic acid, mucilage, wax, glucoside, albumens, starch, etc., and act as astringent, diuretic, tonic and also used in fever and applied to boils and ophthamia. It is an important drug for dissolving kidney and bladder stones.

Bergenia stracheyi (Hook f. & Thoms) Engl.

Family : Saxifragaceae
Trade/Common Name : Pashanbheda

A rhizometic herb with thick rootstock, dark-brown with short stem, thick and

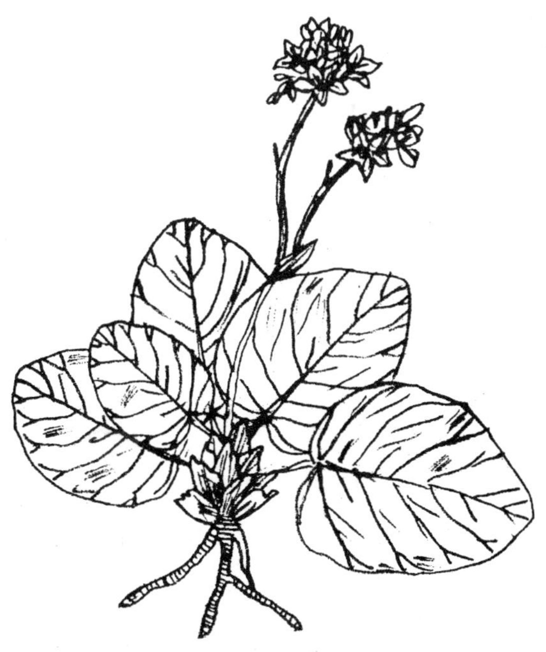

Bergenia ciliata

Bergenia stracheyi

fleshy. Leaves are stalked and sheathing at the base; ovate or orbicular, coradate, margin fringed with short and stiff terminal corymbose cymes and blooming in the month of June-July.

It occurs from Afganistan to Uttar Pradesh between 3300-4500 m in alpine slopes. It is propagated by rhizome cutting.

The rhizome contains gallic acid, tannic acid, mucilage, wax, glucoside, albumens, starch, etc., and act as bitter, astringent, diuretic, aphrodisiac, tonic and also used in fever and applied to boils and ophthamia. Bark is antiseptic, carminative, and is used to heal up the wounds.

Betula utilis D. Don.

Family : Betulaceae
Trade/Common Name : Bhojpatra

A moderate sized, deciduous tree upto 20 m tall or often a mare shrub forming the upper most limit of the vegetation. Bark is smooth, reddish brown, with whitest, linear, horizontal, lenticels. Leaves are stalked, 5-10 cm long, ovate, unequal serrate, acute, slightly hairy, along the mid ribs, pubescent when young and have a rounded base. Male catkins are few, borne at the tips of long shoots; while the female catkins are short, solitary or in pairs and blooming in the month of May-July.

The tree is found in alpine regions of the country between 2900-3900 m altitude. It is propagated by seeds.

Betula utilis

The plant contains betulin, olenolic acid and acetyle olenolic acid in addition to leucocyanadin in the outer bark and polymeric anthocyanidina in the inner bark. The bark is act as acrid, pungent, heating, tonic, alexiteric; useful in convulsions, bronchitis, diseases of blood and the ear, leprosy, etc. The fungal outgrowth, "Bhuja-granthi", is locally used against muscular pains and swellings.

Boerhaavia diffusa Linn.

Family : Nyctaginaceae
Trade/Common Name : Pigweed

A diffuse herb with prostarte or ascending branches. Leaves in unequal

Boerhaavia diffusa

pairs, ovate or suborbicular, mmargins undulate. Flowers pink, minute in small bractate umbels of 4-10, arranged in long-stalked, axillary and terminal panicles and blooming in the nearly throughout the year.

It is found frequently in wastelands and roadsides in suntropical regions below 1000 m elevation. It is propagated by shoot cutting.

The root is diuretic, laxative, stomachic, diaphoretic and anthemintic and used in treatment of jaundice, dropsy, gonorrhoea.

Bombax ceiba Linn.

Syn. *B. malabarica* DC.

Family : Bombacaceae
Trade/Common Name : Silk Cotton Tree

A large deciduous tree. Stem straight, more or less buttressed at the base when old, covered with large Prickles when young; branches whorled, horizontally spreading. Leaves digitate, leaflets 5-7, lanceolate, acuminate, coriaceous, glabrous. Stipules small caduceus. Flowers flashy and blooming in the month of January to March.

It is found throughout India, Burma and Ceylon ascending upto 1000 m or more. It is propagated by seeds.

Bombex ceiba

Gum contains catechu-tannic acid and young roots (Semal-musli) contain proteins, arabinose, glactose, pectose matters, starch and mucilage. The various parts of the plants are used in small pox, bleeding gums, toothache and carries sores in mouth, pain in leg, fever, and enlarged spleen. Atrophy, emaciation, rheumatism, spermatorrhoea, cholera, neuralgia, leprosy etc. Semul-musli (root of young sapling) is stimulant, tonic, and aphrodisiac. Roots and bark are emetic, Gum is known as Mochras, astringent, tonic, haemostatic, aphrodisiac in nature and useful in diarrhoea and dysentery.

Borago officinalis Linn.

Family : Boraginaceae
Trade/Common Name : Common Borago

Borago officinalis

A hardy annual plant upto 60 cm tall, with blue flowers. Flowers are bee shaped, blue and pendulous.

It is native to Europe and Asia. It is cultivated in various parts of India. It is propagated by seeds or cuttings.

It is used for skin diseases, eruptive fevers and phlegmasia. It is also used for flavouring claret-cup and as a bee food.

Bunium persicum (Boiss.) Fedt.

Syn. *Carum bulbocastanum* Clarke

Family : Apiaceae
Trade/Common Name : Kala-jirah

A glabrous herb with globular tuberous roots; profusely branched upto 50-60 cm height. Leaves dissected having narrow segments. Flowers are small, white, in compound umbels and blooming in the month of May-June.

It is native of Baluchistan, Afghanistan and found only in Kashmir and Himachal Pradesh. It is propagated by seeds.

It contains essential oil (2.5%) having carvone (45-65%), a mixture of ketone, carvone, terpene and traces of carvacrol and cuminaldehyde (40%). The seeds are used as stimulant, carminative in the indigenous system and useful in curing diarrhoea, dyspepsia, fever, flatulence, stomachache, hemorrhoids and obstinate hiccup.

Bursera delpechiana Poisso ex Engl.

Family : Burseraceae
Trade/Common Name : Indian lavender

A dioecious plant. It sheds its leaves at the commencement of winter and gives new leaves and flowers in April.

It is found in hilly parts of Indian Himalaya. It is propagated by seeds.

The fruit husk is a source of a very valuable essential oil of commerce which is known as lineo oil or linaloe oil in trade. It is used in perfumery, high quality soaps and cosmetic products.

Butea monosperma (Lamk.) Taub.

Syn. *B. frondosa* Koenig ex Roxb.

Family : Fabaceae
Trade/Common Name : Flame of the Forest

A moderate sized, deciduous tree; bark fibrous, bluish gray or light brown outside, exuding a red juice when cut. Leaves are innately 3-foliolate; leaflet hard, rigidly cariaceous, glabrescent above and silky tomentose beneath. Flowers are in terminal, thick racemes appearing before the leaves on naked branches, giving an appearance of fire flame from a distance and blooming in the month of April-June

Butea monosperma

It is commonly found throughout the greater part of India upto 1300-1400 m except in very arid parts. It is propagated by seeds.

The flower contains butrin, isobutrin, coreopsin and roots contain b-sitosterol, leucoanthocynidin, amyrin, palasonin. Leaves contain glucosides and a yellow tasteless oil. Seeds used as anathematic, aperients, rubefacient and febrifuge. Gum is used in dysentery, diarrhoea. Leaves are used as astringent, tonic; act as aphrodisiac, disperse boils and pimples and are internally used for flatulence and colic, worms and piles. Bark is used as alternative, astringent, aphrodisiac and anathematic. It is also useful in tumor, bleeding piles and ulcers

Callicarpa macrophylla Vahl.

Family : Verbenaceae
Trade/Common Name : Beauty berry

An erect, evergreen shrubs, 2-2.5 m height. Leaves opposite, shortly-stalked, lanceolate, acuminate, crenate or sharply toothed. Upper surface wrinkled pubescent and lower tomentose. Flowers pink, hardly 5 mm long, crowded in axillary, dichotomous cymes and blooming in rainy reason.

It is found in wastelands and ravines areas of sub-tropical regions upto 1200 m elevation. It is propagated by seeds.

Priyangu is an important drug of Ayurveda. It is uses as cooling, astringent, febrifuge and blood purifier and used in the treatment of dysentery, fever and emesis (vomiting). Leaves after heating applied externally on rheumatic pain.

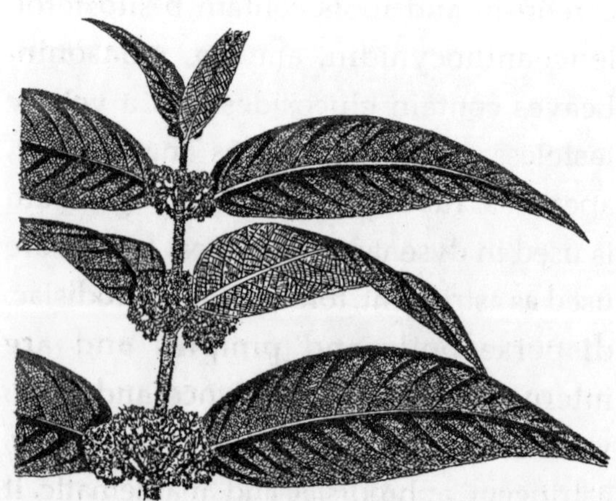

Callicarpa macrophylla

Calotropis gigantea (L.) Ait. f.

Family : Asclepiadaceae
Trade/Common Name : Madar

A small or large shrub, bark ash coloured, wrinkled; leaves are thick, ovate-oblong or broadly oblong, acute covered with white tomentum. Root bark short, corky, deeply vertically cleft pieces. The outer surface is yellowish gray, soft and corky, fissured longitudinally and can be easy separated from middle cortical layer. The flowers are white and blooming throughout the year, but chiefly in the cold season i.e. November-December.

It is found commonly throughout the India in warm and dry places. It is propagated by cutting and seeds.

Leaves and stalks contain calotropin and calotropagenin. Latex contains uscharin

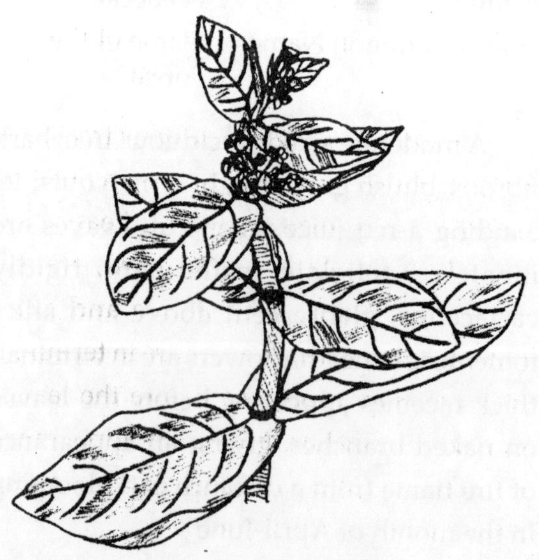

Calotropis gigantea

(0.45%); calotropin (0.15%) and calactin (0.15%). Root bark is useful in dysentery and used as a substitute for Ipecacue; acts as diaphoretics and emetic. Latex is irritant and powdered flowers are useful in cold, cough, asthma and indigestion. Extract of roots and leaves is anti-cancer, used in lupus, tuberculosis leprosy, syphilis, and rheumatism.

Caltha palustris Linn.

Family : Ranunculaceae
Trade/Common Name : Marsh marigold

A perennial herb with shining rounded or heart-shaped leaves with yellow or white cup shaped flowers and blooming in the month of July to continue upto September.

It occurs in marshes, running streams, stony and muddy situations in throughout temperate Himalaya. It is propagated by seeds.

Roots are reported to be poisonous and deaths among horses caused by eating the herb are common. It contains helleborin and veratrin.

Caltha palustris

Canangium odorantum Baill. f.

Syn. *Cananga odorata* Hook. f. & Thorn

Family : Anonaceae
Trade/Common Name : Ylang-Ylang Tree

Canangium odorantum

A large evergreen tree upto 25 m tall. Leaves small, simple alternate. Flowers borne in axillary clusters, long-stalked, drooping with highly fragrant. Flowering from second year onwards. Flowers are collected when turn yellow.

It is native of tropical East Asia. It is cultivated in South India and island of Indian Ocean. It is propagated by seeds or cuttings.

Flowers yield an essential oil (2.0-2.25%) known as Cananga oil. The oil contains sesquiterpenes, sesquiterpene alcohols and esters. The oil has strong and flowery odour. It is used in high grade perfumes. It is also used in soaps and lotions.

Cannabis sativa Linn.

Syn. ***C. indica*** **Lamk**

Family : Cannabiaceae
Trade/Common Name : True Hemp

An annual tall herb upto 5 m. Stem slender, grooved, finely tomentose; leaves are alternate, palmately divided; 5-20 cm long, toothed, rough above, pale downy beneath; leaflet 7-11; flowers dioecious, i.e. borne on different plants; are yellow-green. Male flowers are clustered in short, drooping panicles. Female flowers are sessile, erect and blooming in the month of June-September.

It is widely distributed in the temperate world. It is common on wastelands, field edges, waysides, roadsides and also cultivated in temperate and tropical region particularly in India for narcotic drugs, cordage fibres. It is propagated by seeds.

Flowering and fruiting parts contain resin and volatile oil, cannabine, pseudocannabinol, cannabinin, canin, gum, sugar and potassium nitrate. The Ganja is the dried resinous flowering tops of the female plant free from leaves. Charas is the narcotic, resinous aerial exudation from all the parts of the plant while bhang consists of dried leaves and flowering tops. In action, the drug is anodyne, antispasmodic, anesthetic, aphrodisiac, heating, appetizer, astringent, digestive, exciting, exhilarant, hypnotic, induces sleep, sedative, narcotic and toxic. It is also used to treat depressia, bladder inflammation, gonorrhoea and nervous disorder.

Capparis spinosa Linn.

Family : Capparidaceae
Trade/Common Name : The Caper Bush

A spearding evergreen shrub, low prostrate or trailing in habit with shinning leathery leaves, spinous branches, heavy foliage and white to pink flowers and blooming in the month of July-August.

It occurs in open dry places, silty or sometimes in rocky substartes in Kashmir regions of Himalaya. It also grows in the Western Ghats, Deccan Penninsula, Andhra Pradesh, Rajasthan, Maharashtra, inner valleys of Himalaya and Uttaranchal. It is propagated by seeds.

Capparis spinosa

Flower buds contain rutin, rutic acid, pectic acid and saponin. The root and bark are slightly bitter and used in rheumatism, paralysis, toothache, affections of liver and spleen. It is aperitive, diuretic, resolvent and tonic in nature. The leaf poutice is used to relieve gout pain.

Capsella bursa-pastoris (L.) Medic.

Family : Brassicaceae

Trade/Common Name : Shepherd's Purse

An annual herb upto 60 cm tall, arises from the basal rosette of leaves. Leaves, variable in shape and degree of division in various situations. Flowers are small and white.

Capsella bursa-pastoris

It is native to temperate and subtropical regions. It is propagated by seeds.

The entire herb is medicinally important. It is rich in vitamin C. A decoction is used to treat bloody urine, diarrhoea and haemorrhoids. The juice of the herb is used to clear phlegm from chest, to stop nose bleeding, to increase urine flow and menstrual flow.

Capsicum annum Linn.

Family : Solanaceae

Trade/Common Name : Red Chillies

An annual, much branched herb with ovate, elliptic, pointed leaves. Flowers are

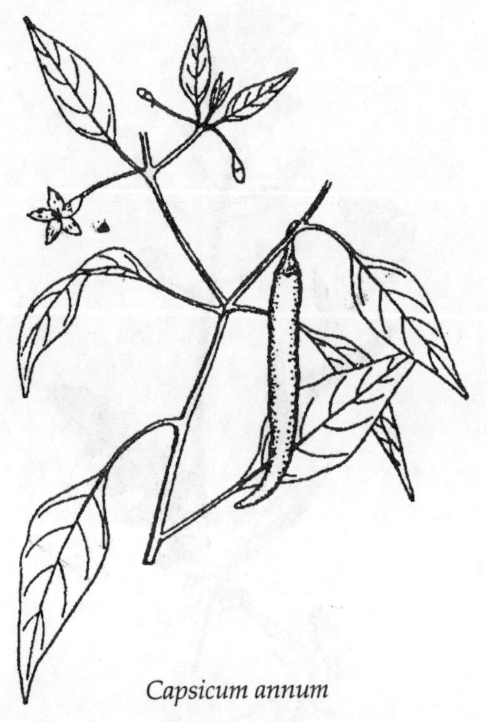

Capsicum annum

white and the fruit, a berry is of variable sizes and shape, red on ripening and many seeded and blooming in the month of May-September.

It is cultivated as kitchen garden crop in throughout India. It is propagated by seeds.

The fruit contains capsicin, capsaicin, solanineand ascorbic acid. It is rich in vitamin A and C, when eaten fresh with salads. It is carminative, digestive, hypnotic, stimulant and stomachic. Useful in diarrhoea, dysentery, filarial, hoarseness, rheumatism, sore-throat, tonsillitis, pharyngitis, neuralgic affections and yellow fever. Green chillies are reported to retard cancer.

Carica papaya Linn.

Family : Caricaceae
Trade/Common Name : Papaya

A perennial herb with milky latex upto 5 m tall. Leaves clustered around apex of the stem and branches. The leaves and unrip fruits contain milky juice having protein-fermeny papain.

It is native to Southern Mexico and Centeral America. It is grown in all tropical countries as well as in some sub-tropical areas. It also cultivated in various parts of India. It is propagated by seeds.

Latex contains papain, chymopapain, amylase, lipase, pectase. It is used to treat eczema, ringworm, warts, corns, calluses, infected wounds, malignant tumers, pain of burns, bleeding haemorrhoids. Locally, small quantity of latex mixed with sugar is

Carica papaya

given for the treatment of enlarged liver and spleen, whooping cough and it is also reharded as emmenagogue and stomachic.

Carthamus tinctorius Linn.

Family : Asteraceae
Trade/Common Name : Safflower

An erect, annual herb upto 1 m tall. Leaves alternate, spiny. Flowers are orange-yellow in colour. Seeds are white and shiny.

It is cultivated throughout greater parts of India. It is propagated by seeds.

Flower and seeds are medicinally important. It is laxative, stimulates menstrual flow and induce perspiration. Flower is used to treat fever and skin eruptions. The herb is also used to dry up the skin symptoms of measles.

Carum carvi Linn.

Syn. *Apium carvi* Linn.: *Seseli carvi* Linn

Family : Apiaceae
Trade/Common Name : Carway

A biennial (or annual) herb, root tapering, brown, often branched below. Stem erect, slender, cylindrical hollow; faintly striate, smooth, much branched; branches ascending. Leaves compound. Flowers white in umbels and blooming in the month of May to July.

Carthamus tinctorius

Carum carvi

The herb, native to Europe and West Asia and now in cultivated in Bihar, Jharkhand, Orissa, Punjab, Bengal, Andhra Pradesh and in the hills of Uttaranchal, Kashmir and Himachal Pradesh. Also found wild in the Northern Himalayan region. It is propagated by seeds.

The fruit contains umbelliferone, scopoletin, herniairin and leaves contain flanoides and glycosides. An essential oil contain carvone (45-65%). As a medicine, it is used as abortifacient, anathematic, cardiac, carminative, deodorant, diuretic, emmenogogue, febrifuge, lactogogue, stimulant, stomachic and tonic. It is useful in ammenorrhoea, blood vomiting with bile protruding piles, rheumatism, intermittent fevers and worm infestation. Locally, poultices of herb is used for swellings in the breast and the testicles.

Cassia absus Linn.

Family : Caesalpiniaceae
Trade/Common Name : Chaksu

An erect, branched viscidly glandular, biennial herb, 15-45 cm tall. Leaves compound, leaflets 4, ovate, 2.5-5.0 cm; obtuse, a small, awl-shaped gland between each pair. Flower few, reddish-yellow, 3-4 cm in diameter appearing in short racemes and blooming in the month of August-September.

It is found in wastelands and stony slopes in sub-Himalayan tract upto 1500 m elevation. It is propagated by seeds.

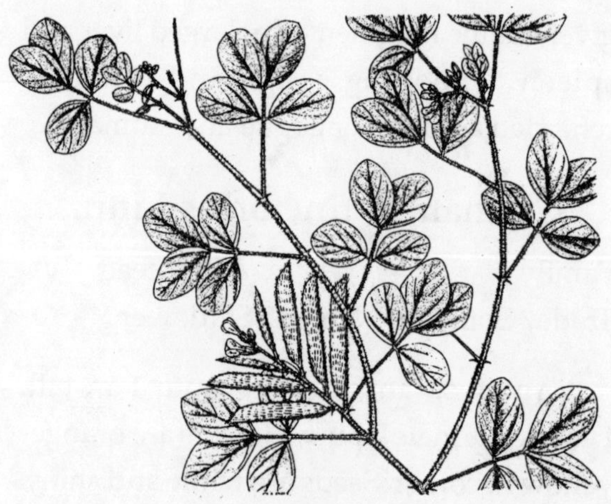

Cassia absus

The roots contain anthraquinones, chrysophanol. Aloe-emodin; alkaloids-chaksine and isochaksine; leaves contain alkaloids quercetin and rutin. Seeds have sitosterol glycosides. Powder of the leaves is used in ulcer dressings. It is one of the constituents of unani drug for leprosy. Seeds are also astringent; used in ringworm and other skin affections; in conjutivities and ophthalmia.

Cassia alata Linn.

Family : Caesalpiniaceae
Trade/Common Name : Ringworm cassia

A large spreading shrub with irregular, angled, glabrous branches. Pods are long, membranous dehiscent with numerous seeds. Flowers are bright yellow in colour

It is found in the various parts fo India. It is propogated by seeds.

The plant contains rhein, anthraquinone

Cassia alata

Cassia angustifolia

derivatives, β-sitosterol and chrysophenol. Leaves are used for curing of ringworm infection and in snake bite. Locally, decoction of the leaves and flowers are used internally in constipation, bronchitis and asthma and also for washing eczematous patches. Roots are diuretic. The juice of the leaves is used as antiherpectic and skin diseases.

Cassia angustifolia Vahl.

Family : Caesalpiniaceae
Trade/Common Name : Indian Senna

A small perennial under shrub with pale suberect or obtusely angled erect or ascending branches. Leaves are large, compound and pinnate, 5-8 jugate, oval-lanceolate, glabrous, bluish green to pale green. Flowers are erect, many with bright yellow and arising on racemes.

It is cultivated as cash crop in south-western tracts of Tamil Nadu in marginal lands. It is propagated by seeds.

The plant contains β-sitosterol (0.33%). The active substrates are the derivatives of anthraquinone. Leaves and pods are cathartic, contain senosides A, B, C, D, rhein and aloe-amine in free and compound forms. Senna leaves are used in the form of infusion. Against the habitual constiputiort, these are used in the form of decoction powder. It is uses as anthelmintic for intestinal worms and as a liver stimulant. Locally, for skin diseases and for removing pimples, a paste of dried leaves mixed with vinegar is used.

Cassia fistula Linn.

Family : Caesalpiniaceae
Trade/Common Name : Indian Laburnum

A medium sized tree upto 15 m tall. Leaves innately compound; leaflet 4-8 pairs, large, ovate-lanceolate, leathery, entire, acute. Flowers are bright yellow in large and drooping racemes appearing like Japanese Lantern and blooming in the month of April to June when leafless.

It is found throughout India upto 1300 m and more common in evergreen forests. The tree is much favored for planting in road side avenues and in gardens. It is propagated by seeds.

The pulp contains rhein, mucilage, protein, and volatile oil. Leaves contain anthraquinone derivaties. Root bark contains tannin, phlobaphenes and oxyanthraquinone substance. The fruit is used as emetic, febrifuge, laxative, reliever of thoracic obstructions and purgative. It is useful in boils, pustules, leprosy, ringworm, colic, gripping, flatulence, dyspepsia, constipation, indigestion, fever and heart diseases.

Cassia occidentalis Linn.

Family : Caesalpiniaceae
Trade/Common Name : Coffee Senna

A diffuse under shrub with loosely spreading branches, often growing 1-2 m

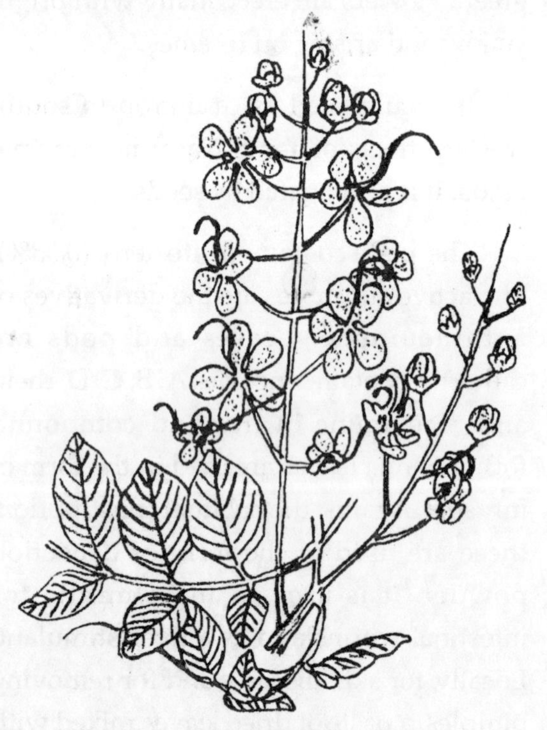

Cassia fistula

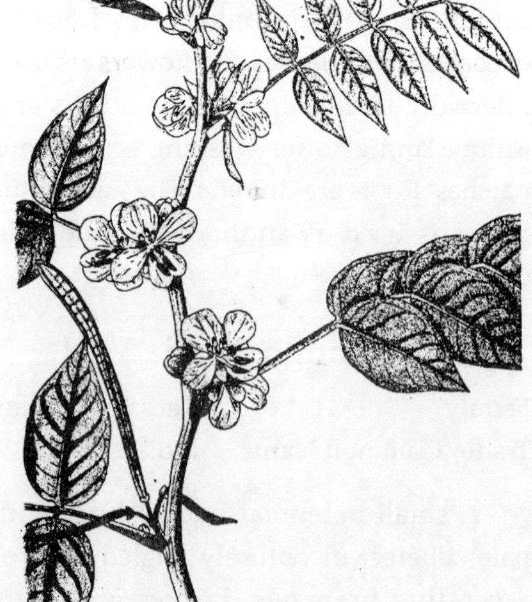

Cassia occidentalis

tall. Leaves pinnate, elliptic-lanceotale, and acuminate. Flowers yellow with reddish veins and blooming in the month of July to August.

A common weed scattered from Himalayas to West Bengal, South India, Burma and Ceylon. It is propagated by seeds.

Seeds contain sennosides, N-Me-morphidine, anthraquinone glucoside; anthraquinones; polysaccharides; galactomannan. Leaves contain dianthronic heteroside. Flower yields phycion and its glycosides, emodin, and β-sitosterol.

It is one of the ingredients of 'Liv-52' used in hepatitis. Root is used in skin diseases. Leaf and fruit are anthelmintic and seeds are used in diabetes.

Cassia tora

Cassia tora Linn.

Syn. *C. obtusifolia* Linn.

Family : Caesalpiniaceae
Trade/Common Name : Sickle Senna

An annual, erect, shrub like herb, 15 cm to 1.5 m tall. Leaves pinnate; leaflets 6, hairy, ovate 2.5-5.0 cm, acute an awl shaped, yellow gland between the lowest pair. Flowers yellow, in pairs on short, maxillary stalks and blooming in the month of August to September.

It is found throughout India, as a weed of waste places, roadsides, field borders etc. in all the warmer places ascending to 1400 m. It is propagated by seeds.

Leaves & stems contain d-manitol, myricyl alcohol and β-sitosterol. Root and seeds yield anthraquinones, β-sitosterol, emodin, subrofusarin, 1,8-dihydroxy anthraquinones etc. The leaves are pounded and applied on cuts and act like tincture of iodine. Seeds with turmeric and mustard oil are applied against eczema. Tender leaves are taken internally to prevent skin diseases while the infusion acts as vermicidal. It has a curative effect in ringwarm in cattle. Seeds powder is useful in fever during teething trouble, itchy eruption, leprosy, psoriasis, tumors, ulcers and hastening of suppuration.

Castanea dentate (Marsh.) Borkh.

Syn. C. *americana* Linn.

Family : Fagaceae
Trade/Common Name : Sweet Chestnut

A hardy decidous tree upto 35 m tall. Leaves are long, margin deeply serrated. 2-3 nuts are found together upto 2.5 cm across, brown in colour, fattened on one side.

It is native to Eastern-North America. It is found in temperate regions of Himalayas. It is propagated by seeds.

Leaves and barks possess medicinal properties. Leaves are used as sedative, tonic, and also treat whooping cough. Bark is used to cure dysentery and worms.

Catheranthus roseus (L.) G. Don.

Syn. *Vinca rosea* Linn.

Family : Apocynaceae
Trade/Common Name : Periwinkle

A perennial branched herb upto 45-120 cm tall. Stem erect, solid, branched. Leaves exstipulate, simple, entire, ovate to elliptic, uni-costate reticulate. Flowers are pink or white, borne in axillary pairs on cymose and blooming throughout the year.

It grows throughout India mostly as escape from garden or as pot herb. It is commercial cultivated in Karnataka, Tamil Nadu and Maharashtra. It is propagated by seeds.

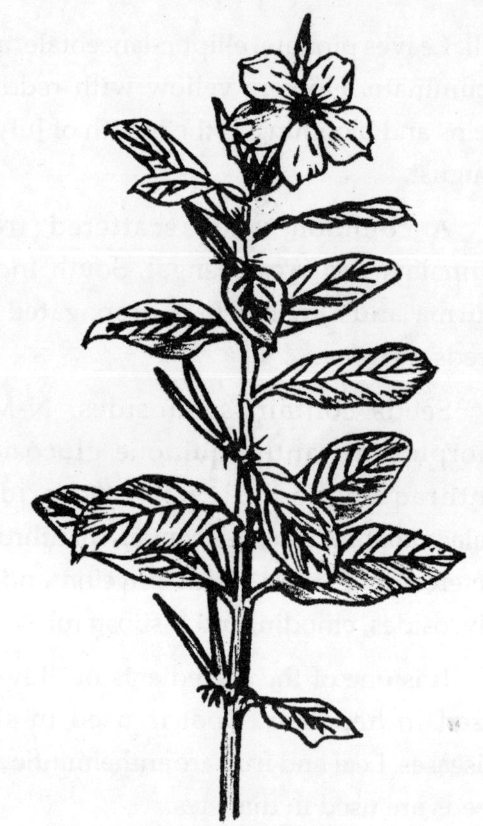

Catheranthus roseus

Roots contain indole alkaloids, raubasin and serpentine. Leaves contain vincristine and vinblastine. It possesses emetic, diaphoretic and antidiabitic properties. Its decoction is effective for regulation of menstruation, for promoting abundant flow of urine in cases of scanty and raddish urine and is also beneficial for kidney. The roots are used to relieve toothache. It has purgative, vermifuge, depurative and haemostatic properties. Its alkaloids are used for the treatment of acute leukaemia in children including lymphocytic leukaemia.

Cedrus deodara (Roxb.) Loud.

Family : Pinaceae
Trade/Common Name : Himalayan Ceder

A large, handsome, evergreen coniferous tree with spreading branches and dark green foliage. Needles arise in tufts at the ends of short branchlets. Wood is aromatic, light, yellowish brown in colour; taste oily and resinous. Male cone cylindrical 5-12 cm erect; female cones ovoid or ovoid-cylindrical, dark brown when ripe while the young cones are bluish purple. Flowers are blooming in the month of September to October.

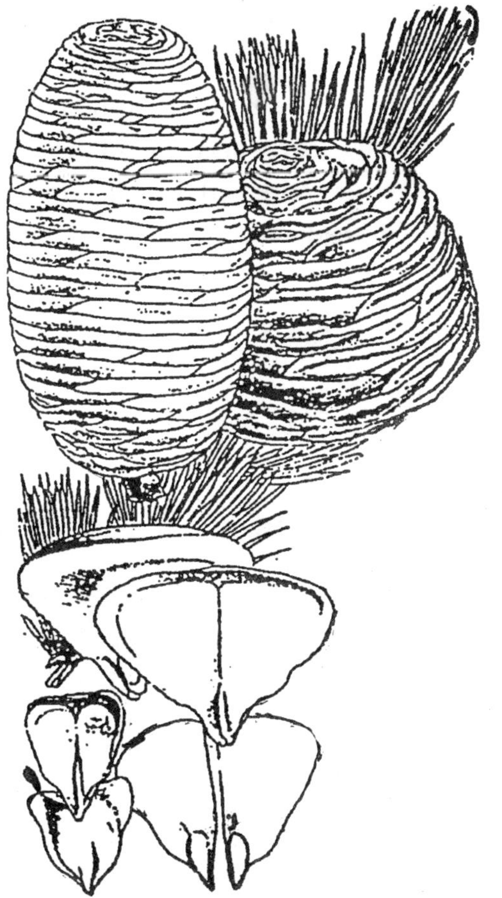

Cedrus deodara

As one of the dominant conifer, it is distributed in most parts of Uttaranchal and Himachal Pradesh between 2200-2800 m. It is propagated by seeds.

The wood oil contains oleo-resin and essential oil while the needles contain ascorbic acid. The wood is carminative, diaphoretic and diuretic; given in fever, flatulence, heart palpitations, paralysis troubles, prolapsed anni, dropsy and in urinary diseases In gonorrhea, syphilis, gout and rheumatism the decoction is given as a powerful alternative with turmeric and guggulu. The tar is used as a favorite alternative and given in chronic skin diseases and in large doses, it is used in leprosy and also applied externally to ulcers.

Celastrus paniculata Willd.

Family : Celastraceae
Trade/Common Name : Jyotismatis

A large straggling shrubs or woody twiners. Branches covered with small and white wart like excrescences. Leaves simple, varible in size and shape, ovate to obovate or orbicular, abruptly pointed, margins toothed. Flowers yellowish-green, in terminal, drooping panicles and blooming in the month of April to continue upto July.

It is found in sporadic in scrub forests and along roadsides in sub-tropical regions below 1200 m elevation.

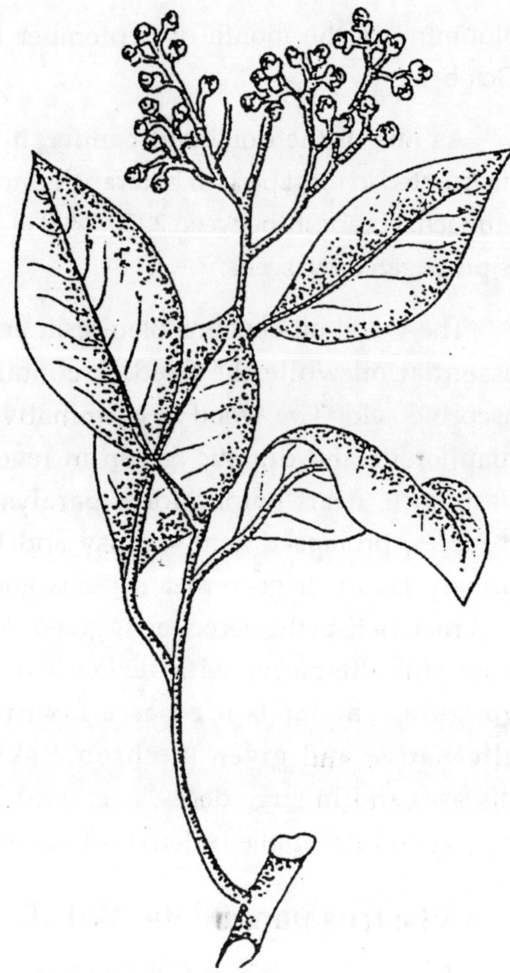

Celastrus paniculata

Seeds are used as stimulant, laxative and aphrodisiac. The seed oil is useful in rheumatism and also in strengthening memory.

Centella asiatica (Linn.) Urban.

Syn. *Hydrocotyle asiatica* Linn.

Family : Apiaceae
Trade/Common Name : Mandukparni

The herb trails on ground and its creeping stems bear roots on their nodes. Leaves are long stalked, 2-4 cm in diameter, sometimes more, rounded to broad or kidney-shaped, cordately based, deeply toothed margins. Flowers are minute, pinkish and red, 3-6 in cluster and blooming in the month of July-August.

It grows in moist places throughout India. It is commonly seen in marshy banks of rivers, steams, ponds, in irrigated lawns and field etc. below 1300 m in elevation. It is propagated by stolen.

The herb contains bitter substance, aglucoside-asiaticoside, essential and fatty oil, sitosterol. Tannin and resinous substance besides an alkaloid hydrocotylin. A bitter principle vallarine, pectin acid and resin are present in the leaves and roots of the plants. The leaves or entire herb is boiled in water and this decoction is given in the treatment

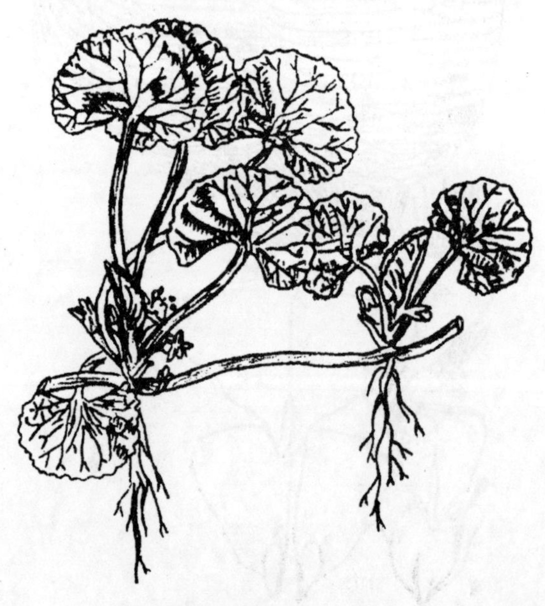

Centella asiatica

of leprosy and useful in epilepsy and is a tonic for brain. It has also been reputed to improve the power of speech and poetic imagination. It is a useful remedy for mentally retarded children.

Centratherum anthelmiticum (Willd) Kunz

Syn. *Vernonia anthethmintica* **(L.) Willd.**

Family : Asteraceae
Trade/Common Name : Somraj

An erect tall herb. Leaves long, covered with minute hairs and margins toothed. Flower heads are upto 2.5 cm cross, purplish appear in small clustered of 30 to 40 minute flowers.

Centratherum anthelmiticum

It is found throughout India mostly in waste places near habitations upto 1400 m elevation. It is propagated by seeds.

Seeds are medicinally important. It is stimulant, antiseptic and anthelmintic. It destroys worms and promote urination. It is also used to treat skin diseases.

Cephaelis ipecacuanha A. Rich.

Family : Rubiaceae
Trade/Common Name : Ipecac

A small, perennial herb upto 75 cm tall. Stem slender, cylindrical, trailing, with few or no branches. Leaves opposite pairs near the top of the plant, alternate, pointed tips, margin entire, roughly hairy. Flowers white, funneliform, sessile, arises in dense clusters.

Cephaelis ipecacuanha

It is native to Bolivia and Brazil. It is commercial cultivated in Darjeeling and parts of Assam and Sikkim and also in South India. It is propagated by seeds, root and leaf cutting.

The root contains 2.2 to 2.5% total alkaloids. The principal are emetine and cephaeline including psychotrine methyl ether, emetamine, ipecacuanic acid ipeoside. It is used as an expectorant, emetic, diaphoretic, antiamoebic agent and vomitive. It is useful in treatment of haemorrhages, amoebic, dysentery, acute bronchitis, whooping cough. In small doses, it stimulates apetite and improves digestion.

Cestrum nocturnum Linn.

Family : Solanaceae
Trade/Common Name : Queen of Night

Cestrum nocturnum

A quick growing bushy, evergreen, subscandent shrub upto 2 m height. Branches are glabrous, weak and yellowish. Leaves alternate, lanceolate, ovate, oblong, pointed at the apex. Flowers are yellow or greenish yellow, grow in axillary on terminal panicles in clusters and highly scented in night, blooming during summer and rainy season.

It is native to India. It is found in several parts of India. It is propagated by seeds or cuttings.

Flowers on distillation yield 130 ppm of oil and flower concrete yield is around 0.26%. It is used in selective perfumes and cosmetics.

Chenopodium album Linn.

Family : Chenopodiaceae
Trade/Common Name : Wild spinach

A hardy annual herb upto 1.2 m tall. with whitish, fuzzy surface. The lower leaf is whiter than the upper surface. Flowers are arising in spikes from the upper leaf axils and greenish in colour.

It is found as a weed in the roadsides, in fields, waste lands etc in throughout India. It is propagated by seeds.

Leaves are rich in vitamin C and midly laxative. Locally, tea of the leaves is used to relieve stomach pains.

Chenopodium album

Chenopodium ambrosioides Linn.

Family : Chenopodiaceae
Trade/Common Name : American wormseed

An erect, aromatic, much branched herb, upto 100 cm tall. The fruits are somewhat globular, slightly compressed with a thin pericarp surrounding the seeds. The entire plant possesses a comphoraceous odour. A volatile oil of medicinal value is found in the glandular hairs, especially of the pericarp of the fruit. The flowers are minute and clustered in pale green leafy spikes and blooming in the month of July to August.

It is native of West Indies and South America, the plant is widely distributed in waste places, roadsides and field borders in South India, Bengal, Maharashtra, Uttaranchal, Himachal Pradesh and Kashmir. It is propagated by seeds.

The leaves contain kaempferol-7 shamnoside and ambroide. The pericarp contains an essential oil, calcium, phosphorus, oxalates etc. The plant is an anthelmintic and its oil is used in medicines. The oil is effective against many forms of

Chenopodium ambrosioides

intestinal parasites and also useful in the treatment of amoebic dysentery.

Chlorophytum arundinaceum Baker.

Family : Liliaceae
Trade/Common Name : Safed Musli

A rhizomatous herb. Leaves are suberect, lanceolate aand many nerved. Seeds black with angular edges. Infloresence is an erect, dence-flowered racemes. Flowers are star like, white upto 2 cm across.

In India, it is widely distributed in tarai regions of Bihar and West Bengal and also in some parts of Orissa, Madhya Pradesh and Chhatisgarh. It si propagated by rhizomes.

Dry roots possess less than 5% moisture and also contains carbohydrate, proteins, root fibres, saponins and minerals. It possess tonic, aphrodisiac properties and used to treat general debility. Locally, tubers (25 g) is boiled with milk and is taken twice a day for a month in cases of importency and weakness.

Chlorophytum borivilianum San. & Fer.

Family : Liliaceae
Trade/Common Name : Safed Musli

A herbaceous plant with a cluster of fleshy roots. Root fibres cylindric, leaves 6-18, oblanceolate, obtuse acute or acuminate.

Chlorophytum borivilianum

It is found in Sal forest and moist teak forests of Madhya Pradesh, Chhatisgarh, Jharkhand etc. It is propagated by seeds.

Root stock contains rich in glycosides. It is used for rheumatic and nervous complaints and also useful in diabetes.

Chondrodendron tomentosum Ruiz.

Family : Menisperma-ceae
Trade/Common Name : Pareira

A climbering woody vine. Stem warty, bark velvety grey. Leaves are heart shaped or nearly round, upper surface slightly hairly, and lower surface woolly white. Male and female flowers borbe on separate plants.

It is widely distributed in Asian countries upto 600 m altitute. It is propagated by seeds.

Roots and stem contain quaternary base d-tubocurarine and crystalline bases. They possess febrifuge properties and used to treat rheumatism, jaundice, leucorrhoea, gonorhoea, kindney and bladder inflammation. It is a muscle relaxant and also acts as relaxant in neurological cases.

Chonemorpha macrophylla G. Don.

Syn. C. *fragrans* Alston.

Family : Apocynaceae
Trade/Common Name : Wild Thorn

An attractive evergreen heavy climber. Stem is pubescent with white latex. Leaves are opposite, orbicular or obovate, large upto 25 cm long. Flowers appear in terminal cymes in clusters from May to June, creamy white, sweet scented.

It is native to India. It is found in moist forest of Himalayan regions. It is propagated by seeds or layering.

Chrysanthemum balsamita Linn.

Family : Asteraceae
Trade/Common Name : Mint Geranium

A coarse perennial herb upto 90 cm tall. Leaves are entire, silvery-hairy, crenate, petioles are as long as blade, possess balsamic scent. Clusters of bright yellow flowers are borne during August.

It is native to Europe and West Asia. It is cultivated throughout India as ornamental. It is propagated by seeds or cuttings and divisions.

Its ray florets are used medicinal purposes.

Cichorium intybus Linn.

Family : Asteraceae
Trade/Common Name : Wild Endive

A tall herb, 30-120 cm high, having grooved steak and wide spreading branches with long fleshy taproot. Leaves are oblong-lanceolate, basal leaves palmately lobed, cauline, entire, bract like, clasping. Flower heads bright blue, solitary, terminal, often clustered and maxillary flowers are arranged along stout, erect, green, nearly leafless, spreading ray florets and blooming in the month of July to August.

It is native of the temperate parts of the old world and it found in the Punjab, North-West parts and Hyderabad. It has naturalized in Sonmarg area in Kashmir. It is propagated by seeds.

The ash of the root is rich in potash. The bitter principle is the glycoside of fructose and pyrocatechuic acid. The juice of roots contain stearin, mannite and tartaric acid.

Cichorium intybus

The herb is useful in fever, vomiting, diarrhea, enlargement of spleen. The root is good stomachic and diuretic. The wild form is considered tonic and alexiteric.

Cinchona officinalis Linn.

Family : Rubiaceae
Trade/Common Name : Indian Cinchona

It is distributed in Nilgiris and Annamalai of Tamil Nadu and Darjeeling hills of West Bengal.

The bark contains 6-7% alkaloids and major ones are quinine and quinidine. It is

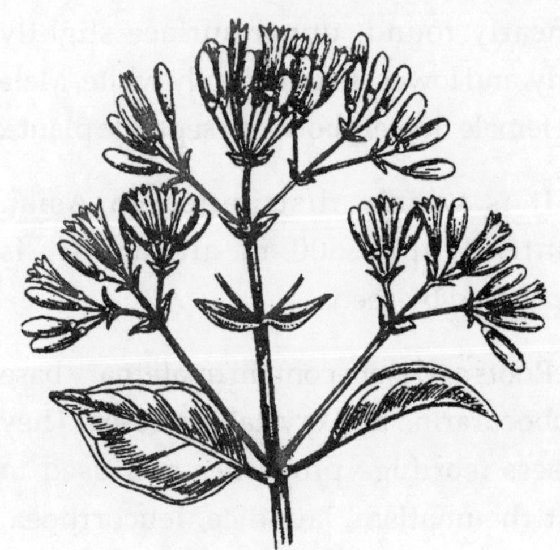

Cinchona officinalis

antiseptic, acting mainly on the infusoria. These alkaloids are used as a curative to malaria, also as anasthetic, protected skin against sunburn and stomachic appetizer. It is also employed as antiarrhuthmic agent and used as treatment of heart disease.

Cinnamomum camphora (L.) Nees & Ebesm.

Family : Lauraceae
Trade/Common Name : Camphor tree

A medium size tree with an enlarged base and reaches a height of about 12 m. Leaves smooth, shiny, whitish beneth upto 12 cm long with strong camphor smell. Flowers are yellow in colour.

It is native to China and Japan and also cultivated in Nilgiri hills of India. It is propagated by seeds and cuttings.

Leaves containing an essential oil and used to treat diarrhoea, rheumatism, bronchitis, pneumonia and muscular pains.

Cinnamomum camphora

It also stimulate uterus, menstruation and uterine haemorrhages.

It contains tannin, it has an astringent effect, increases constipation unless mixed with laxative. It is known to be antiseptic, circulatory stimulant and has a calming effect in cases of hysteria, neuralgia and general nervousness.

Cinnamomum tamala Nees & Ebesm.

Family : Lauraceae
Trade/Common Name : Tejpat

A small to medium sized evergreen tree, 2-10 m tall. Leaves are stalked, opposite, or sub-opposite, elliptic-oblong, 3-nerved from the base, shining, leathery, entire, long pointed. New leaves are slightly pinkish

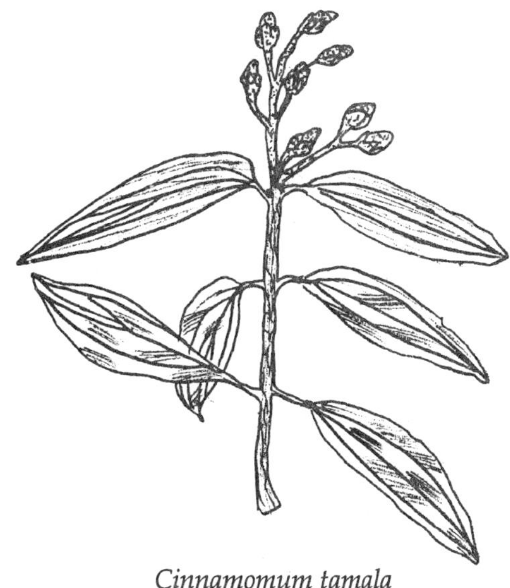

Cinnamomum tamala

tinged. Flowers are small, yellowish and blooming in the month of March to May.

It is found in Sub-Himalaya, Khasi and Jaintia hills. It is cultivated in certain parts of India. It is present sporadic in some parts of Himachal Pradesh and Uttaranchal. It is propagated by seeds.

The herb yield in essential oil. The bark is used in gonorrhoea and leaves used as stimulant, carminative and useful in colic, diarrhoea, rheumatism etc.

Cinnamomum zeylanicum Blume.

Family : Lauraceae
Trade/Common Name : Ceylon Cinnamon.

An evergreen tree, attains a height of 15 m. Leaves are shiny green, above and white on the underside, opposite, oblong-ovate, bluntly pointed at the tips with a

Cinnamomum zeylanicum

spicy odour. Bark smooth, thick and scabrous, ash coloured. Flowers are yellowish-white in colour.

It is native to India. It is cultivated in southern India. It is propagated by seeds or shoot cuttings.

Cinnamon chips yield an volatile oil (0.10-0.20%). It contains methyl-n-amyl ketone, furfural, p-cymene. Its leaves also yield an oil (0.25-0.50%). The main constituent of the leaf oil is eugenol which about 90% of the oil. Its oil is highly priced and most flavouring ingredients for baked goods, candies, soft drinks, table sauce and all kinds of confectionaries. It is a powerful bactericide.

Cirrhopetalum maculosum Linn.

Family : Orchidaceae
Trade/Common Name : Wild Blue Orchid

An epiphytic orchid with rhizomes, covered with fibrous sheaths and many roots. Pseudobulbs are ovoid and shriveled. Leaves are long, linear, notched. Scape stout, to 7 cm long, umbels 2-5 flowers with pale yellow green colour.

It is native to Uttaranchal between 1000 to 1400 m altitude. It is propagated by rhizome cuttings.

Roots are medicinally important and used to increase longevity.

Cissampelos pareira Linn.

Family : Menisperma-ceae
Trade/Common Name : Ice vine

A twining, perennial herb. Leaves are alternate, globular, cordate, kidney shaped, soft, hairy. Flower are greenish, dioecious i.e. male and female are borne on different plants. Male inflorescence in a raceme and blooming in the month of April to June.

It is found frequently in hedges and over shrubs, along waysides throughout India upto 1500 m elevation. It is propagated by seeds.

The root contains an alkaloid isoquinoline, pelosine, berberine and saponin. Also contain respine, cissampeline, hyatin, hyatinin, a quercitel and a sterol. The roots are used as anthelmintic, antidote to poison (including snake poison), antilithic, astringent, cardiac, carminative, diuretic, expectorant, febrifuge sedative, suppurative

Cissampelos pareira

and tonic. It is also useful in asthma, cold & cough, colic, diarrhoea and dysentery, fever, indigestion, inflammation of kidney and bladder, piles and ulcers.

Cissus quadrangularis Linn.

Syn. *Vitis quadrangularis* Wall.

Family : Vitaceae
Trade/Common Name : Hatjora.

A tendril climber, stem quadrangular, green with distinct nodes and inter nodes, tendrilsare at nodes. Leaves opposite.

It is found in warmer parts of India. It is propagated by cuttings.

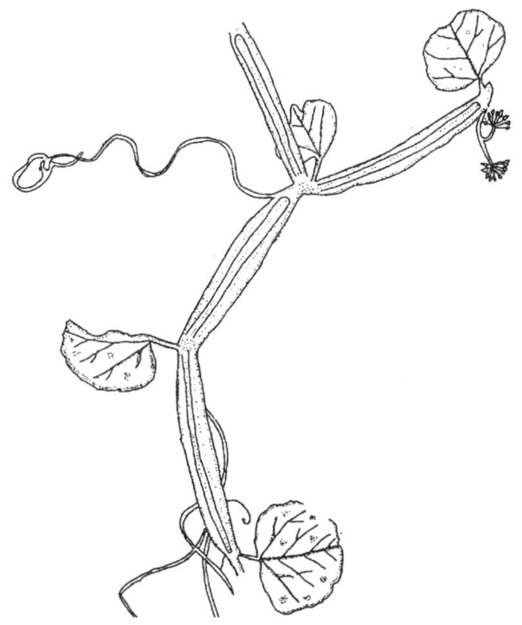

Cissus quandrangularis

It contains calcium oxalate, ascorbic acid. The leaves and young shoots are considered as stomachic, alternative and used in powder form in digestion. The juice of the stem is useful in irregular menstruation and scurvy. The stem given internally and applied topically for fracture of bones; beaten into a paste given for asthma.

Citrus aurantifolia (Christm.) Swingle.

Family : Rutaceae
Trade/Common Name : Sour Lime.

A much branched spiny shrub or tree. Leaflets are elliptic-ovate, up to 7.5 cm long. Petioles are narrowly winged. Flowers borne in racemose clusters, white and fragrant.

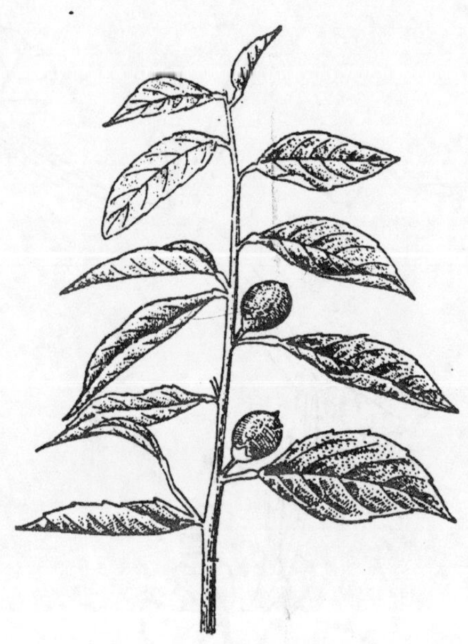

Citrus aurantifolia

It is wild in certain parts of India and also cultivated in the plains, and up to 1500 m altitude for its fruits. It is propagated by seeds, budding or grafting.

Fruit yields an oil (0.28-0.40%). It contains L-pinene, furfural, B-pinene, n-octyl aldehyde, borneol, citral, geraniol, terpineol. Lime oil is the chief flavouring ingredient of carbonated, non alcoholic beverages, cola drinks, ice-creams, confectionaries, candies and baked goods.

Citrus aurantium Linn.

Family : Rutaceae

Trade/Common Name : Bergamot Orange.

A tree grows to a height of 4 m. Leaves are deep green, long and glabrate. Flowers are white in colour, fragrant and small. Fruit peel is thin and tough with numerous oil glands.

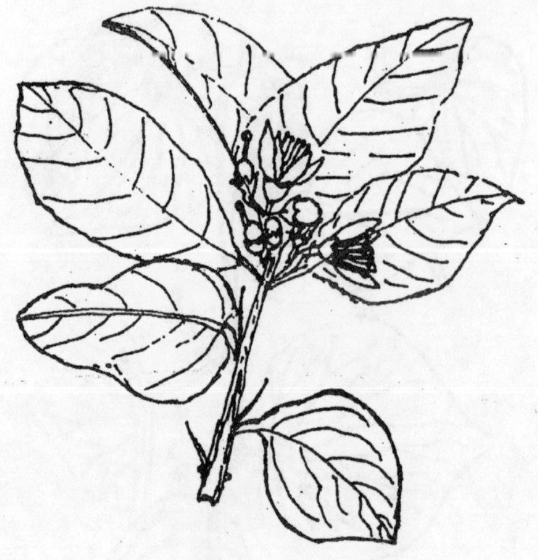

Citrus aurantium

It occurs in some parts of India. It is propagated by seeds or grafting.

Leaves and fruit peel yield an essential oil (0.4-0.5%). The principal constituent of the oil is linalyl acetate; and others are octylene, alpha-pinene, camphene, l-linalool, geraniol, bergamotlin. Leaves of bergamot orange yield 0.33-0.40% of oil. The peel oil is one of the important fine perfume material and its refreshing scent blends with any perfume composition. Its largely used in soaps, powders, creams, lotions etc.

Citrus lemon (Linn.) Burm.f.

Family : Rutaceae

Trade/Common Name : Lemon

A small tree of 6 m tall with short and stout spines. Leaflets are oblong to elliptic-ovate, up to 10 cm long. Petioles are not winged. Flowers are 5 cm across, reddish in bud stage. Fruits are ovate or oblong, up to 12.5 cm long long with a terminal nipple yellow in colour and very acidic in taste.

Citrus lemon

It is found throughout India. It is propagated by seeds or grafting or cuttings.

Rinds of fruits, flowers and leaves contain essential oil. The citral content in lemon oil varies from 2.0 to 3.7%. It contains alpha-pinene, beta-pinene, d-limonene, linalool, aldehyde, acetic acid etc. Lemon peel oil is an important

Flavouring agent and widely used in soft drinks, soft drink powders, tablets, beverages, baked goods, cakes, pastries, confectionery, candies, ice-creams and gelatin.

Citrus sinensis (Linn.) Osbeck.

Family : Rutaceae

Trade/Common Name : Sweet Orange

A compact tree of 12 m tall with stout spines. Leaflets are 10 cm long, elliptic to oblong-ovate. Petioles are narrowly winged. Flowers are borne in clusters of 1 to 6. Fruit is globose, orange with smooth skin and solid core.

Citrus sinensis

It is widely grown in sun tropical regions of India. It is propagated by seeds, grafting or cuttings.

Fruit peels yield an essential oil. It contains d-limenene, decyl aldehyde, citral, octyl alcohal, linalool forming acid and capric acid. The leaves, leaflets and twigs oil contain pinene, limonene, dipentene, aliphatic hydrocarbons, nerol, linalyl acetate, geranic acid, etc. Sweet orange peel oil is used for flavouring of food products, beverages and confectionaries. For pharmaceutical and oral preparations in perfumes, creams, soaps, powders, lip sticks and toilet preparation.

Claviceps purpurea (Fr.) Tul.

Family : Clavicipitaceae
Trade/Common Name : Ergot of Rye

A parasitic fungus exits on the heads of rye (*Secale cereale* L. Family- Ascomycetes). Mycelia of the fungus when come in contact, grow in the ovaries of the flowers of the host and form fruit like bodies called sclerotia. This sclerotis is a horn like hard structure, dark purple or nearly black in colour.

It is found in throughout India. It is propagated spores.

Ergot sclerotia contains principal alkaloid ergoline. Others are ergonovine, ergotanine, ergocryptine, ergosine and ergovalide. It stimulate the smooth muscles of uterus including rhythmic contractions. Its alkaloids are used for controlling hypertension and peripheral vascular disorders.

Clematis nepaulensis DC.

Family : Ranunculacaeae
Trade/Common Name : Ghantiali

A perennial, hardy, glabrous climber. Lower sides of the petiole sensitive to contact. The petiole bends once round the support thickens and lignifies. Flowers are greenish in colour.

It is found moist shady areas of temperate zones between 1500 to 2400 m altitude in Uttaranchal Himalaya. It is propagated by cuttings.

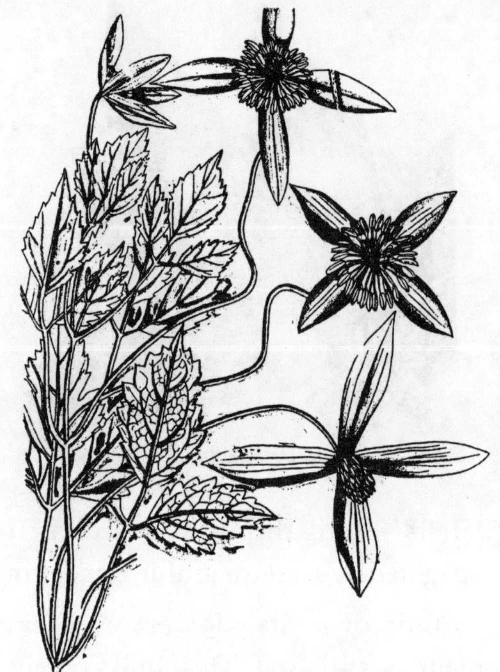

Clematis nepaulensis

The herb is acrid and poisonous due to presence of anemonin. Its leaves are used as irritating agent, harmful to skin.

Clerodendrum serratum Spreng.

Family : Verbenaceae
Trade/Common Name : Bhargi

A perennial shrub, 1-2 m high, with bluntly four angular stems. Leaves in whorls, sessile, narrow lanceolate, sub-entire, glabrous, rather hard. Flowers are bluish-purple often white, in pyramid shaped terminal panicles and blooming in the month of July to August.

It is found in Deccan and Carnatic, West-Coast district of Tamil Nadu, Uttaranchal, Himachal Pradesh; from Sikkim and Assam

Clerodendrum serratum

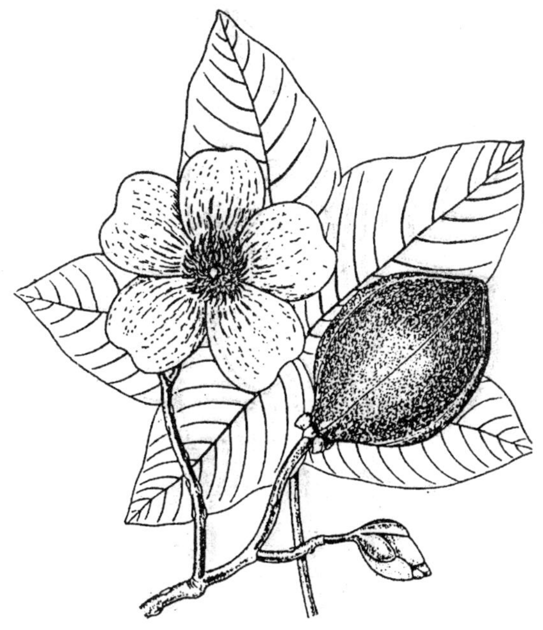

Cochlospermum gossypium

to Tenasserium and cultivated for ornamental purposes. It is propagated by cutting and seeds.

The leaves contain alkaloid and bitter substance. The root contains sapogenin and triterpene, starch and resin. The roots are useful in asthma, cough and scrofulous affections; loss of appetite, dropsy and fever. The resin and leaves are useful in syphilis, rheumatism and act as febrifuge and bitter tonic.

Cochlospermum gossypium DC.

Family : Cochlospermaceae
Trade/Common Name : Golden Silk Cotton Tree

A deciduous tree or shrub up to 10 m tall. Leaves crowed near the ends of branches, palmately 3 5 lobed, glabrous above, whitish-grey tomentose beneath. Flowers greenish-yellow, appear before the leaves, in terminal panicles.

It is native to India and found in dry forests. It is propagated by seeds or cuttings.

Gum is mild demulcent used in coughs. Dried leaves and flowers are used as stimulant. The bark infusion of about 20 ml is taken in jaundice twice a day for fifteen days.

Codonopsis ovata Benth.

Family : Campanulaceae
Trade/Common Name : Sardandi

A hardy, ascending to twining herb with thick white, milky tuberous roots, often large, long and thick enough as compared

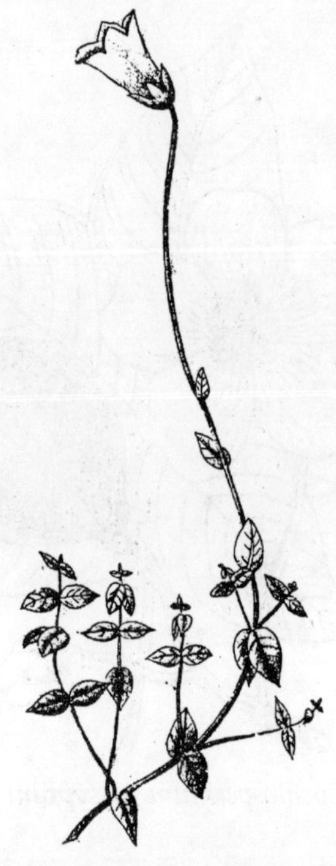

Codonopsis ovata

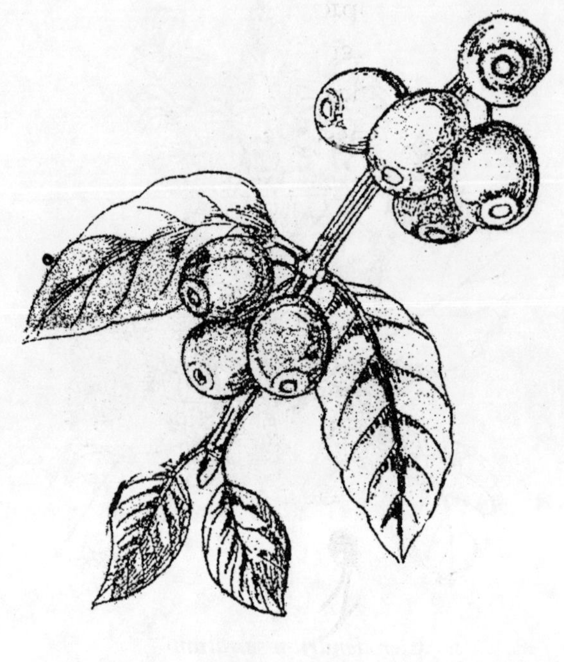

Coffea arabica

to the weak stems. Flowers are cup-shaped, greenish yellow with purplish streak, appearing during July-September.

It is native to Western Himalaya. It is found sporadically along with alpine ferns between 3500-3800 m of Himachal Pradesh. It is propagated by seeds and cuttings.

The roots are considered as aphrodisiac and stimulant.

Coffea arabica Linn.

Family : Rubiaceae
Trade/Common Name : Coffee

A small evergreen tree upto 5 m tall. Leaves are glossy, oblong-elliptic, opposite, pointed at the tip, upto 15 cm long. Flowers are showy, fragrant, abundant, white in colour, clustered in the leaf axils.

It is native to Ethiopia and commercially cultivated in various parts of India. It is propagated by seeds and also by vegetative means.

Seeds containd an active principal Caffeine (1-3%). It promotes formation of pepsin and hydrochloric acid in gastric juice. It is a stimulant and facilitates digestion after a heavy meal. It is a diuretic and reduces cutaneous perspiration. It is useful in dynamic typhoid fever, intermittent fever, periodic nervous asthma, whooping cough and albuminuria.

Colchicum autumnale Linn.

Family : Liliaceae
Trade/Common Name : Meadow Saffron

A perennial herb upto 30 cm tall. In the spring leaves appear and the capsules is brought above ground by lengthening its stalk. The corm is conical or ovate, fleshy with numerous fibrous roots. In autumn the flower projects out of the soil. Flowers are lavender to light pink.

It is distributed in the Netherlands, Denmark, England, Ireland, Poland, Spain, Italy and North Africa. It is propagated by seeds.

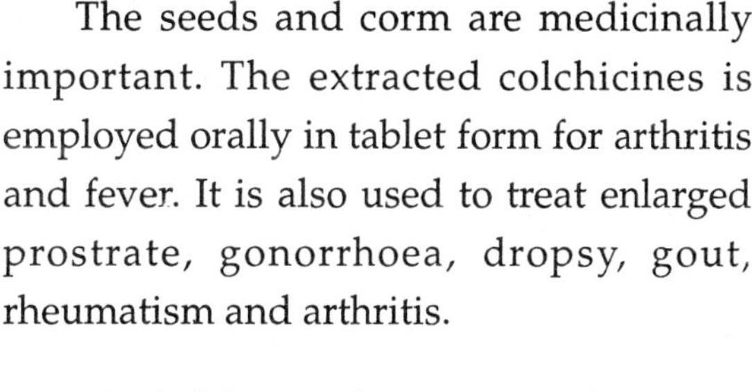

The seeds and corm are medicinally important. The extracted colchicines is employed orally in tablet form for arthritis and fever. It is also used to treat enlarged prostrate, gonorrhoea, dropsy, gout, rheumatism and arthritis.

Colchicum luteum Baker.

Family : Liliaceae
Trade/Common Name : Virkim Posh

A perennial herb, Corm cylindrical, 15 cm deep in soil, Leaves appear after the flowering, Flowers coming out of soil with the melting of snow and blooming in the month of July-August.

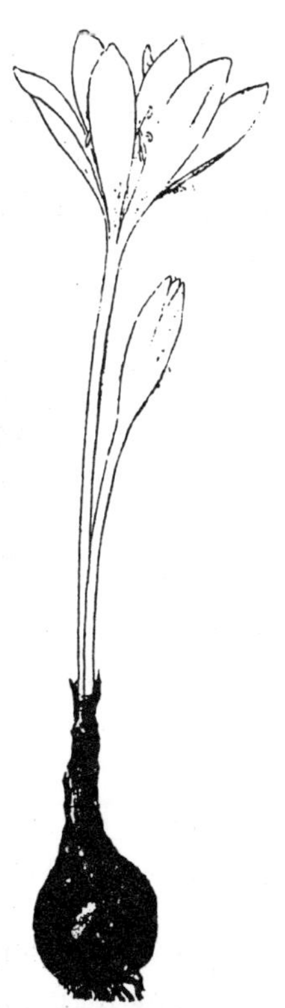

Colchicum autumnale

Colchicum luteum

It occcurs in open moist slopes, preference clay loam in Western temperate Himalaya. It is propagted by corms.

Corm (dry) contains colchicine (0.21-0.25%) –an alkaloid used to induce polyploidy in plants. Seeds contain 0.41-0.45% colchicine. The seeds and corm possess carminative, laxative, aphrodisiac, alternative, and aperient and useful in rheumatism and diseases of liver and spleen. Locally, the water extract/poútice of corm is used to relieve pain of gout patients.

Coleus forskohlii (Willd.) Briq.

Syn. *C. barbatus* Benth.

Family : Lamiaceae
Trade/Common Name : Indian Coleus

A annual or perennial, erect, pungent-aromatic herb with stem much branched from the woody base, patently hairy up to 60 cm high. Leaves ovate or obovate, glandular-hairy. Flower are purple or pale-blue in colour, appear in terminal spike-like racemes and blooming in the month of August-September.

It is found in dry barren hilly areas of India. Now, cultivated in some parts of South Gujrat, Rajasthan, Maharashtra, Karnataka and Tamil Nadu. It is propagated by seeds.

The tuberous roots contain forskolin and developed as drugs for hypertension, glaucoma, asthma, congestive heart failures and certain types of cancers. Its foliage is employed in treatment intestinal disorders and used as a condiment since long time.

Coleus forskohlii

Commiphora mukul (Hook. ex. Stocks.) Engl.

Syn. *C. wighti* (Arn) Bhand; *C. roxburghii* (Arn) Engl.

Family : Burseraceae
Trade/Common Name : Guggal

A shrub with corked, knotty branches ends in sharp spines upto 4 m height. Leaves 1-3 foliolate, alternate or fasicled, leaflets are glabous.

It is found in the tracts of Western India. It is propagated by seeds and root cuttings.

It is a source of Indian Bdellium, an oleo-gum-resin obtained by incision of the bark. It is used as incense, as a fixative in

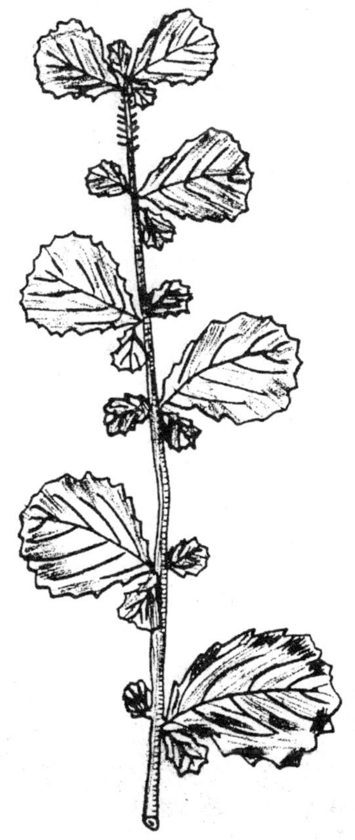

Commiphora mukul

perfumery and in medicine. In indigenous medicine, it is used as an astrigent, antiseptic, stomachic, carminative and digestant. Due to its property of lowering the cholesterol level of the blood it is good demand in modern medicine also. Its oleo-gum-resin is used in the treatment of aryhritis and obesity.

Coptis teeta Wall.

Family : Ranunculaceae
Trade/Common Name : Goldthread

A hardy evergreen bog plant and a small stemless herb. Leaves are compound and glabrous. Flowers are small, regular,

Coptis teeta

white in colour. Fruits are many seedsed follicles and seeds black.

It is native to Himalayan region and distributed in temperate regions of Arunachal Pradesh. It is propagated by seeds or divisions of roots.

The dried rhizomes are medicinally important and contains berberine alkaloid. It is a tonic and stomachic and used for the treatment of fevers and debility.

Coriandrum sativum Linn.

Family : Apiaceae
Trade/Common Name : Coriander

An annual, well branched herb, upto 90 cm tall. The lower leaves are broader with

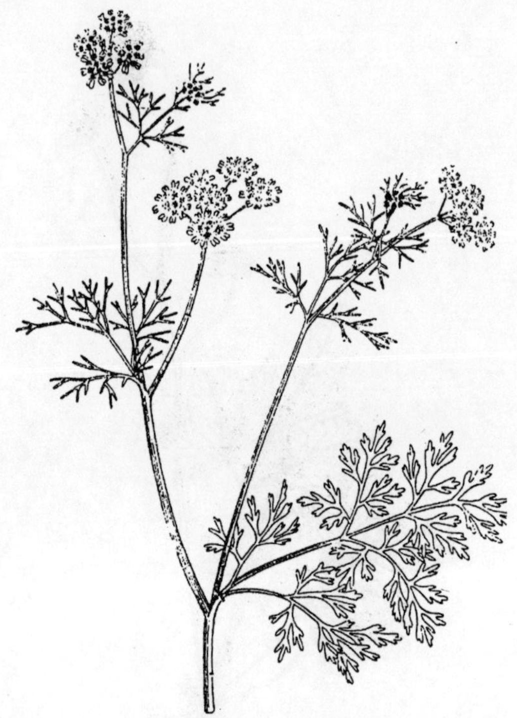

Coriandrum sativum

crenately lobed margins, while the upper ones are narrow, finely cut with linear lobes. Flowers are small, white, often pinkish-purple tinged in terminal compound umbels and blooming in the month of February to March.

It is cultivated throughout India for its seeds and leaves as a culinary herb and a condiment. It is propagated by seeds.

The fruit contains umbelliferone, scopoletin, rutin, coumarins, quercetin, kaempferal and aflatoxins. The seeds contain an essential oil with geranyl contant. The fruits are used as condiment in confectionery and used as anthelmintic, aphrodisiac, aromatic, astringent, carminative, digestive, diuretic, stimulant and tonic. It is also useful in burning sensation of the body; in colic, chronic conjunctives, dropsy, diarrhea, dysentery, dyspepsia, headache, as a heart tonic.

Corydalis govaniana Wall.

Family : Fumariaceae
Trade/Common Name : Bhutbishi

A soft, slender, herb with long, thick, woody, finely knit type, dark brown rootstock, crowned with withered leaf sheaths. Radical leaves are long stalked, 2-pinnate, cut to the base and as long as

Corydalis govaniana

stem. Leaflets-wedge shaped and deeply lobed. Flowers are yellow, crowned, 2.5 cm long and arising from 5-10 cm long individual axis and blooming in the month of June-September.

It is widely distributed from Pakistan to E. Nepal, 2400-4800 m in shrub berries, open slopes of Alpine region. It is propagated by seeds and cutting of underground stems.

The roots contain a variety of alkaloids including protopine and stylopine. The roots are considered as a tonic, diuretic, alternative and antiperiodic. It is prescribed in syphilitic, scrofulous and cutaneous affections.

Coscinium feniestratum (Gaertn.) Colebr.

Family : Menispermaceae

Trade/Common Name : Jharihaldi

A woody climber. Leaves are leathery, roundish, hairy on lower surface. Bark is yellowish. Flowers are in small heads, minute and green.

It is native to Indomalayan region. It is found in western Ghats. It is propagated by seeds.

Dried stem contains berberine alkaloid. Bark is antiseptic and roots have antibiotic properties. The dried stem is used to treat dyspepsia and general weakness.

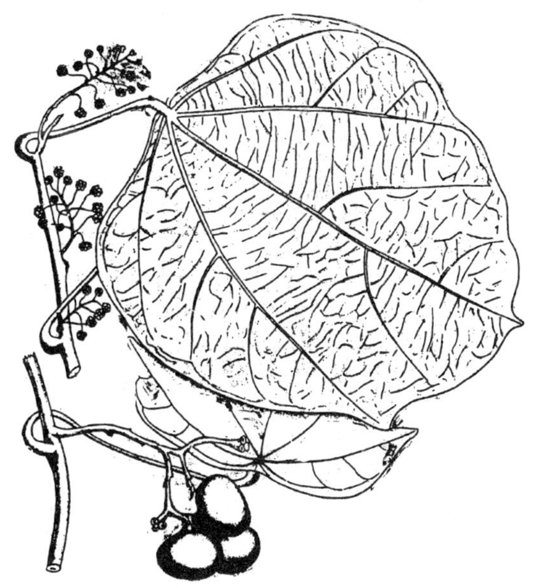

Coscinium feniestratum

Costus speciosus Smith

Family : Zingiberaceae

Trade/Common Name : Costus

An erect stout, rhizomatic, perennial herb; subsessile, spirally arranged, oblong, acute or acuminate leaves; glabrous above and silky pubescent beneath. Flowers white, funnel shaped with contrasting, large, bright red bracts, numerous in a very dense cylindrical spike 5-12 cm long by 3.8-7.5 cm in diameter borne on a stout red leafly stem ranging from 1-2 m. Flowers are orange red, rounded forming a funnel with margins incurved and meeting and blooming in the month of June-July.

It is found more or less throughout India upto 1500 m elevation in forest and shrubberies. It is propagated by seeds.

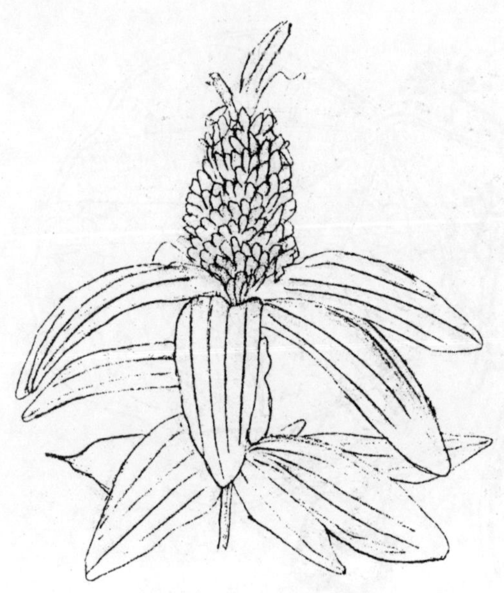

Costus speciosus

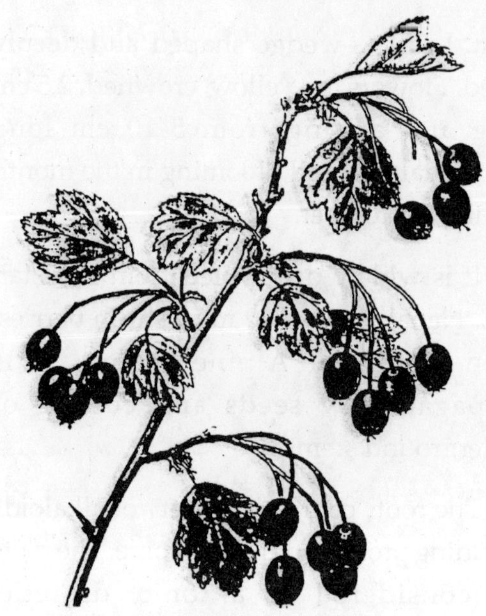

Crataegus oxyacantha

The rhizome contains cortico-steroids and starch. It is bitter, astringent, anthelmintic, used in snakebite; useful in bronchitis, fever, Kapha and Vata, dyspepsia, anemia, rheumatism, hiccough and inflammation. It is used to treat cough, catarrhal fever, dyspepsia, skin disease, worms and snake bite.

Crataegus oxyacantha Linn.

Family : Rosaceae
Trade/Common Name : Hawthorn

A deciduous small tree upto 8 m in height. Leaves deeply cut to 2/3 their width, blade broadly ovate or rhombic in outline. Flowers are white in lax-topped clusters and blooming in the month of July to continue upto September.

Globally, it is distributed from Murree to Kishtwar, Afghanistan, Westward to Atlantic, W. Siberia. In India, it occurs in woodlands, scrubs foothills, cultivated places, wastelands in Western temperate Himalaya. It is propagated by seeds.

The dried fruits possess astringent, cyanogenetic, depurative, diuretic, fumitory, stomachic properties. It is useful in the treatment of cardiac disorders of nervous origin. It is used as home remedy for asteriosclerosis, dropsy, hypertension and sore throat. Locally, it is used for making herbal tea.

Crataeva nurvala Buch-Ham.

Family : Capparaceae
Trade/Common Name : Thorn Wing

A small medium sized tree. Leaves are digitatively trifoliat

Crataeva nurvala

It is native to India. It is found throughout the greater parts of India. It is propagated by seeds.

Bark contains tannins and saponin. It is used as bitter, anti-periodic, tonic and demulcent. It is useful in calculus and affections of urinary organs.

Crocus sativus Linn.

Family : Iridaceae
Trade/Common Name : Saffron

A bulbous, perennial, herb with long, linear, channeled leaves. Corm of walnut size with fibrous scales remaining upto 30 cm deep in soil. Flowers deep violet, funnel shaped, stalkless, and appearing in autumn with the leaves and blooming in the month of October to November.

The herb is native of South Europe and West Asia, cultivated in dry raised slopes in Kashmir valley, especially in pampore area upto an altitude of 1600 m. It is propagated by bulbs.

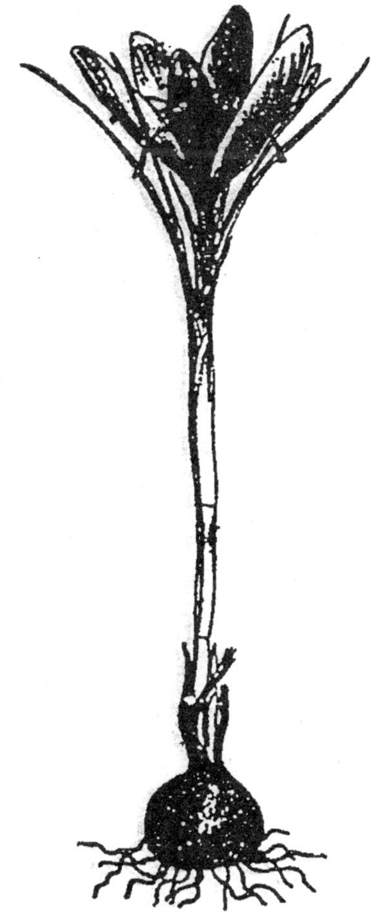

Crocus sativus

The herb contains glucoside, crocin, crocetin, picro-crocin essential oil (0.8%), also contain carotene and lycopene, riboflavin and thiamine. The saffron is used principally for its colouring and flavoring properties. It is used colouring butter, cheese, puddings, pastry and confectionery and also used as a nerve sedative, stimulant, stomachic, abortifacient and as emmenagogue and is used in fevers and snake bite, melancholia and the enlargement of liver and spleen and is a good remedy for catarrhal affections of children.

Croton tiglium Linn.

Family : Euphorbiaceae
Trade/Common Name : Purging Croton

The plant possess stellate hairs and scales.

It is distributed in throughout India as a ornamental plants. It is propagated by seeds and shoot cuttings.

Seeds yield an essential oil. It contains oleic acid, linoleic, palmitic, stearic. It is powerful purgative, irritant and vesicant and is also a co-carcinogen.

Croton tiglium

Cupressus arizonica Greene.

Family : Cupressaceae
Trade/Common Name : Cypres Tree

A hardy evergreen coniferous tree upto 12 m tall. Leaves are small scale like. Outer bark is thin and dark brownish with vertical fissures.

It is found in temperate regions of Himalayas. It is propagated by seeds or by cutting.

Leaves are used to treat ulcers, tumors, toothache, skin inflammation and ringworm. Stem bark are applied to damaged skin and burns.

Curculigo orchioides Gaertn.

Family : Amaryllidaceae
Trade/Common Name : Kali Musli

Curculigo orchioides

A stemless herb with tuberous rootstocks. Leaves are fleshy, elongate, lorate or ensiform.

It is native to India and distributed in Peninsular India, Khasi hills and subtropical Himalayas. It is propagated by sucker.

It is used as a substitute of safed musli. Rhizome used in gonorrhoea, jaundice, piles and in diarrhoea, also used as an aphrodisiac.

Curcuma angustifolia Roxb.

Family : Zingiberaceae
Trade/Common Name : East Indian arrow root

Curcuma angustifolia

A stemless herb with tuberous root stocks bearing sessile and long fleshly fibes terminating in pale oblong pendulous tubers. Leaves with petiole, 30-45 cm blade lanceolate acute 15-30 cm long. Flowering spikes lateral apart from and usually appearing earlier than the leafy spike, crowned by several enlarged empty pink bracts.

It is found moist deciduous Sal and mixed forest of Madhya Pradesh, Chhatisgarh and Jharkhand. It is propagated by rhizome cuttings.

Root stock is used as tonic, useful in leprosy, burning sensations, dyspepsia, loss of taste, bronchitis, asthma, jaundice, anaemia, leucoderma, stones in kidney, urinary discharge, ulcers and diseases of blood.

Curcuma caesia Roxb.

Family : Zingiberaceae
Trade/Common Name : Black Turmeric

A stemless herbaceous plant with tuberous root stock bearing sessile and long stipitate tubers. Leaves usually 30-60 cm by 12.5-15 cm broadly lanceolate or oblong. Petiole and sheath long. The dark blue colour of the mid rib along the one end to the other on upper surface of the leaves is prominently found. Spike appears with peduncle rather before or at the time of leafing in the month of May to early June. Flowers are pale yellow, reddish at the lower border, rather shorter than their bracts.

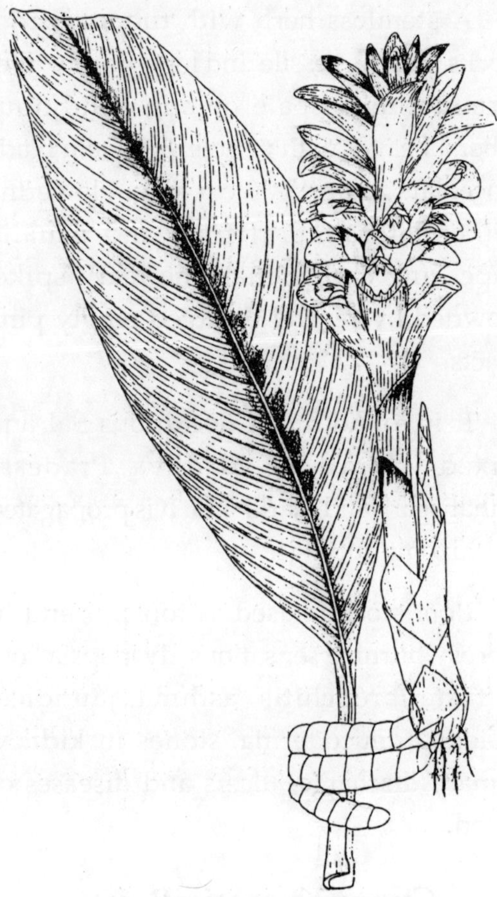

Curcuma caesia

It is found rarely in sal forest of Madhya Pradesh and Chhatisgarh. It is propagated by root cuttings.

Its rhizome is pungent, bitter, fragrant, destroys foul asthma. It is useful in leucoderma, piles, tomours, tuberculosis glands of the neck and enlargement of the spleen etc. The fresh root considered to be cooling and diuretic, it checks lecorrhoeal and gonorrhoel discharge and purifies blood. The juice of the leaves is given in dropsy.

Curcuma domestica Val.

Syn. *C. longa* Linn

Family : Zingiberaceae

Trade/Common Name : Turmeric

A feshy rooted perennial herb upto 60 cm long. Stem short with tufled leaves. Flowers are yellow.

It is native to India and China. It is extensively cultivated in India, China, Indo-China and Sri Lanka. In India, it commercial grown in Andhra Pradesh, Maharashtra, Tamil Nadu, Karnataka, Kerala and Orissa. It is propagated by rhizomes.

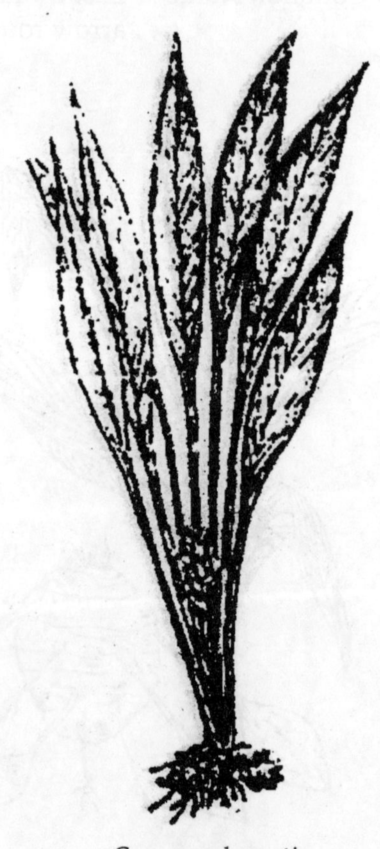

Curcuma domestica

It contains protain, fat, carbohydrate, fibres, vitamin A, B, C, and niacin. Rhizome yields an essential oil (5-6%). It is a spice crop used in industries and medicines. It is stomachic, blood purifier, vermicide, antiseptic, carminative and tonic. It is prescribed as an anti-periodic, alternative, in case of diabetes and leprosy. Its powder is considered sacred and used as cosmetic by Hindu women.

Cuscuta reflexa Roxb.

Family : Convolvulaceae
Trade/Common Name : Amarbel

A leafless, twinning/spreading parasite throwing profuse branches and often covers the host shrubs or other plants completely. Stems/branches are fleshy; flower small, white, pedicel led and arranged is loses racemes or often in short clusters and blooming in the month of April to May.

It is found throughout India and Ceylon and grows upto 2800 m elevation in the various parts of Himachal Pradesh. It is propagated by stem cutting.

The plant contains amerbelin, cussutin and wax. The herb is used internally in protracted fevers, as astringent and anthelmintic. Its decoction made with aromatic is given in flatulence, induration in liver, diarrhoea. The seeds are carminative and purgative. Powered seeds are used as an antifertility drug. The juice of the stems and branches is used to kill lice.

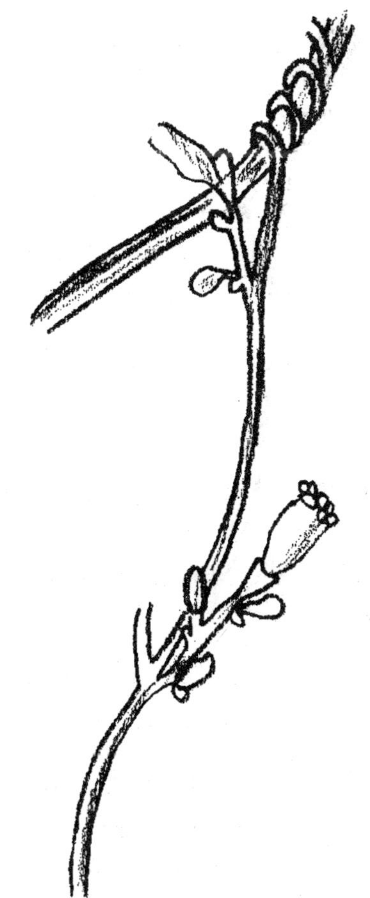

Cuscuta reflexa

Cyamopsis tetragonoloba (Linn.) Taub.

Syn. *C. psoralioides* DC.

Family : Fabaceae
Trade/Common Name : Clustered bean

An erect herb with appressed, laterally attached hairs. Flowers in axillary racemes, small and purplish.

It is native to India and Pakistan. It is cultivated as fodder in various parts of India. It is propagated by seeds.

Gum contains hydrogalactose and anhydromannose. It is used as a binder and employed as suspensions or thicker in emulsions.

Cymbopogon flexuosus (Nees ex Steu) Wat.

Family : Poaceae
Trade/Common Name : Cochin or Malabar grass

A herbaceuous grass with long, narrow and upright leaves with light of medium green colour. Flowering season is dry season (March-April).

It is cultivated in South India especially in Kerala. It is propagated by slips.

The leaves yield 03.-0.5% (total oil) from fresh grass. Out of total oil contains myrcene (0.02%), p-cymene (0.04%), terpinene (0.6%), b-terpineol (0.4%), a-terpineol (2.25%), triacetate (0.9%), borneol (1.9%), geraniol + nerol (1.5%), citrol (85-90%), farnesol 7.2-7.4), etc. Ionone prepared from citrol present in oil is one of the most important raw material of vitamin 'A'. Its oil is mainly used in the manufacture of perfumes for soaps, hair oils, scents and medicine. It also has the antibacterial properties, Ionone prepared from citrol is used for the preparation of vitamin A. In addition to its use in perfumery, Ionone is used in certain kinds of confectionery and liquors. It is also used to improve the flavour of fish, wines and sauces etc.

Cymbopogon martinii (Roxb.) Wats

Family : Poaceae
Trade/Common Name : Palmarosa grass

A tufted perennial grass with numerous stiff stems, yellow or greenish-yellow in colour. Leaves are long, narrow, lanceolate to linear-lanceolate

It is grown in M.P., Chhatisgarh, Maharshtra, A. P., Assam, etc. It is propagated by seeds or slips.

Leaves yield an essntial oil (0.15-0.40%). It contains geraniol (95%), citronellol (2%) and linalool (4%). The oil is used extensively in perfumery, especially in soaps. It is also used for flavouring tobacco and certain pharmaceutical industries.

Cymbopogon martinii

Cymbopogon winterianus Jowitt.

Family : Poaceae
Trade/Common Name : Citronella grass

A herbaceuous grass.

It is grown mainly in Assam and to a limited extent in states like U.P., Uttaranchal, A. P. , Karnataka, Gujrat and Tamil Nadu. It is propagated by slips.

Cymbopogon winterianus

Leaves yield an oil is known as Citronela oil (1.0-1.2%). It contains citronellal (32-45%), geraniol (12-18%), citronellol (11-15%), geranyl acetate (3-8%) etc. The oil is used in aroma chemicals such as citronellal, hydroxy citronellal, synthetic menthol, esters of geraniol etc. It is also used for scenting soaps, in the manufacture of deodorants and mosquito repellent creams.

Cynodon dactylon (Linn.) Pers.

Family : Poaceae
Trade/Common Name : Doob-grass

A creeping herb, rooting at the joints and sending smooth, upward stems. The roots are whitish, tough and creeping, almost woody with smooth fibres. Leaves tapering to a sharp point, ribbed, with smooth sheath and hairy stipules. Flowers, purplish, arranged in 2, close, alternate rows in equally crowed 4 or 5 terminal, linear spikes and blooming in the month of August to September.

The plant is distributed and grows almost everywhere. It is propagated by stolon.

The herb contains β-sitosterol, β-carotene, vitamin-C, palmitic acid, triterpenoids, alkaloids-ergonovine, erogonovivine; ferulic, syringic, vanillin, p-coumaric acid; furfural, alcohol, glucose,

Cynodon dactylon

fructose etc. Decoction of the root is used as diuretic, in dropsy, in hysteria, epilepsy, insanity; astringent and in secondary syphilis. Infusion of root is useful in stopping the bleeding from piles. The juice of the plant is astringent, used as application to fresh cuts and wounds.

Cynoglossum glochidiatum Wall. ex. Benth.

Family : Boraginaceae
Trade/Common Name : Indian Hound's Tongue

A hardy perennial, hairy bulbous herb. Leaves and stem consists of white hairs. Flowers are blue.

It is native to Himalaya, widely distributed in the barren lands, road sides and grass lands between 1200-3500 m in Himachal Pradesh. It is propagated by divisions of bulbs and seeds.

The purplish brown juice is applied eye drops in conjunctivitis.

Cyperus rotundus Linn.

Family : Cyperaceae
Trade/Common Name : Nut grass

A troublesome weed in rice fields. Culms are erect, triangular, with rounded angle. Leaves-radical, sheathing, shorter than the culms. Roots-tuberous, sweet smelling, rounded black and hard like a knot; often crowded together. Tubers are black externally, white internally and smell like *Acorus*. Umbels are terminal, compound, 3-leaved and blooming in the month of July to August.

It is found throughout India in moist area, rice-field and along water courses upto 1500 m elevation. It is propagated by seeds.

The tubers contain terpenes; olealonic acid and its glycosides; β-sitosterol and essential oil. The essential oil contains pinene, traces of cineole, monoterpens, sesquiterpenes and patchoulenone. The roots is used as stomachic, tonic, diuretic, demulent, stimulant and galactagogue; given in fevers, dyspepsia, diarrhoea and cholera; also in urinary calculi and amenorrhoea. It is useful in appeasing of thrist, disorders of stomach, irritation of bowels, febrile and dyspeptic affections; heals wounds and ulcers; pain in the abdomen and in scorpion stings.

Cytisus scoparius Link.

Syn. *Sarothamnus scoparius* (L.) Wim. ex. Koch.

Family : Fabaceae
Trade/Common Name : Common Broom

A hardy perennial shrub upto 3 m tall. Flowers are yellow in colour. Pods are brownish black, villous only in margin.

It is native to Europe and Western Asia. It is found in Nilgiris, Himachal Pradesh and elsewhere. It is propagated by seeds, cutting and grafting.

Dried plant tips contains sparteine alkaloid and the flavone scoparin. It possess cardiac, depressant, cathartic, diuretic and oxytocic properties and used to induce and stimulate labour.

Dactylorhiza hatagirea (D. Don.) Soo.

Syn. *Orchis latifolia* Auct. Non. L.

Family : Orchidaceae
Trade/Common Name : Salampanja

The plant is terrestial orchid in alpine meadows with fleshy tuberous roots. The tubers are slightly flattened, palmately lobed like a hand with fingers open, hence the name 'Panja'. Stem is usually 30-50 cm tall, leafly and with a few sheathing scales in the lower portion; Leaves are erect, oblong-lanceolate, obtuse and with a sheathing base, Flowers are rosy-purple to dull purple in colour, crowded in terminal dense, spicate racemes are spurred and the darker lip is spotted and blooming in the month of June to July.

The plant is found in shrubberies, open slopes, marshes places in the Western temperate Himalayas from Himachal Pradesh to Uttaranchal at 2800-4200 m in elevation. It is propagated by division of tubers.

The roots contain starch, fairly large quantity of mucilage, sugar, albumen, a trace of a volatile oil and ash consisting chiefly of potassium and lime. The roots are used as a farinaceous food, nervine tonic and aphrodisiac, mucilage jelly is nutritious and useful in diarrhoea, dysentery and chronic fevers, demulcent, nutritive, astringent and is given in paralytic affections etc.

Dactylorhiza hatagirea

Datura fastuosa Linn.

Family : Solanaceae
Trade/Common Name : Thorn Apple

A robust, erect, 60-120 cm tall herb having woody stem at the base and purplish

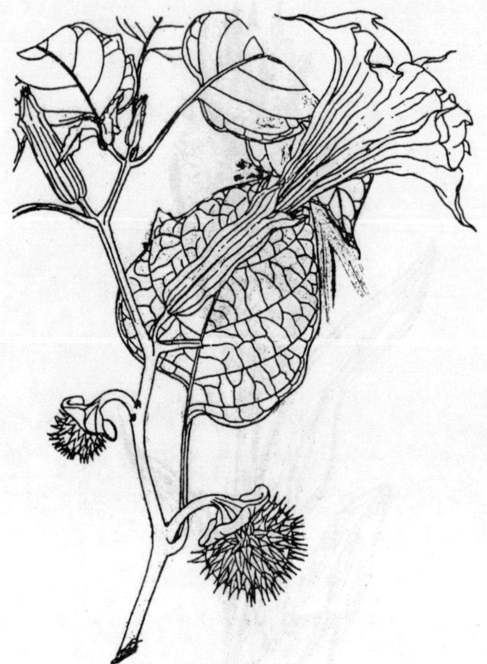

Datura fastuosa

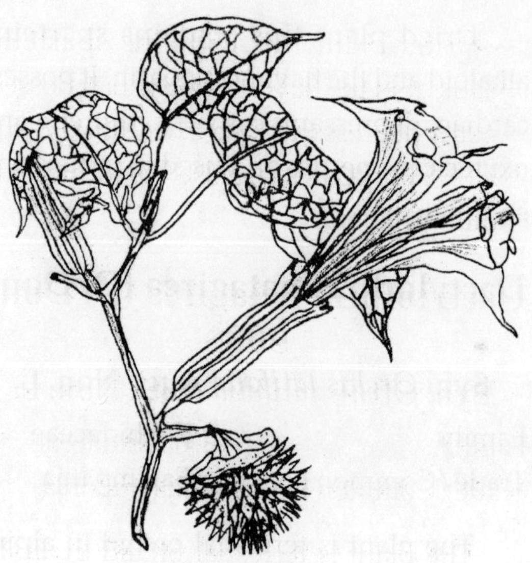

Datura innoxia

to dark-purplish upwards. Leaves long, broad, ovate or ovate-lanceolate acute or acuminate, sinuate, dentate with cuneate base, unequal sided; Flowers pedicilate, solitary and blooming in the month of May to August.

It is sporadically in Shivalik belt areas on waste grounds, roadsides and debris dumping sites. It is propagated by seeds.

The leaves & seeds contain scopolamine, hyoscine and hysocyamine. It is smoked to cure coughs. Seeds and making liquors more intoxicating. Roots are used to cure toothache and for brushing teeth.

Datura innoxia Mill.

Family : Solanaceae
Trade/Common Name : Dhatura

An annual, bushy hern upto 1.2 m tall. Corolla of its flowers is 10 angled broad. Fruits have very slender prickled. Seeds are brown.

It is found in Bangalore, Ahmedabad, Pune, Lucknow, Oilani and Jammu. It is propagated by seeds.

The dried leaves contain alkaloids chiefly as hyoscyamine with smaller amount of hyoscine and atropine. Leaves contain vitamin C & D. The drug is narcotic and anodyne. It is useful in dysmenorrhoea, neuralgia and sciatica and in boils, breast inflammation caused by excessive formation of milk; asthma, salivation, travel sickness, diarrhoea, enlarged testicles, insanity and itch.

Datura metel Linn.

Family : Solanaceae
Trade/Common Name : Black Dhatura

Datura metel

The herbaceous whole plant densely clothed with grayish tomentum. Stem erect, stout, herbaceous, terete. Leaves long, broadly ovate, acute, unequal at the base and often cordate, densely tomentose on both surfaces and generally glandular. Flower pale-white, large, erect and blooming in the month of April continues upto August.

It is found as a weed of waste places in allover India, occasionally grown in gardens. It is propagated by seeds.

The plant contains scopolamine, hyoscyamine, hyoscine, atropine and vit. C. The herb is violently narcotic. Leaves and roots area good palliative in spasmodic asthma. Seeds, leaves and roots are considered useful in insanity, fever with catarrhal and cerebral complications, diarrhoea, skin diseases, lice infestation etc.

Datura stramonium Linn.

Family : Solanaceae
Trade/Common Name : Mad Apple

An annual, erect, nearly glabrous, tall, robust herb with broad, ovate, coarsely and irregularly lobed and toothed leaves. Flowers are white, single or short, usually axillary stalk and blooming in the month of March continues upto August.

It occurs in temperate and sub-tropical parts of Northern Hemisphere, but naturalized in the Himalaya, upto 2700 m, in cultivated areas, waste lands and near village habitations. It is propagated by seeds.

The dried leaves and flowering tops contain 0.3- 0.5% of alkaloids chiefly as hyoscyamine with smaller amount of hyoscine and atropine. The dry seeds contain 0.033% and fruits 0.038% of total alkaloid of which 25% is scopolamine and 75% hyoscyamine. Leaves contain vitamin

Datura stramonium

C & D. The drug is narcotic and anodyne. It affects the sympathetic nervous system, but does not affect the motor or the sensory nerves. Its effect is similar to that of atropine. It is useful in dysmenorrhoea, neuralgia and sciatica and in boils, breast inflammation caused by excessive formation of milk; asthma, salivation, travel sickness, diarrhoea, enlarged testicles, insanity and itch.

Daucus carota Linn.

Family : Apiacaeae
Trade/Common Name : Wild Carrot

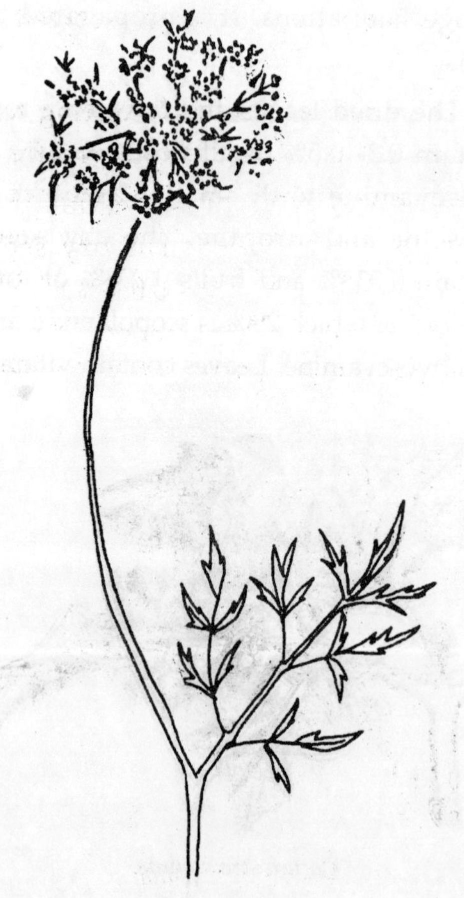

Daucus carota

An erect, hardy, branched, fuzzy, bennial herb with thick fleshy root. Leaves pinnately divided. Flowers are borne in cluster. In center of the umbel is usually a red terminal flower. Fruit oblong with some bristles.

It is native to Asia and Europe and found in various parts of country. It is propagated by seeds.

It contains high carotene, vitamin A and carbohydrate. Fruits, seeds and roots are medicinally important. It is used to increase urine flow, stimulate menstruation, intestinal worms. An appetizing drink is prepared from purple and black carrots. Locally, its boiled roots is applied as poultice on cuts and bruises.

Delphinium denudatum Wall. ex Hk. f. & T.

Family : Ranunculaceae
Trade/Common Name : Nirbishi

A annual herb upto 70 cm in height. Leaves with rounded blade, cut into 3-5 narrow lobes. Follicles hairless. Flowers are blue or violate borne in a widely branched infloresence with a few terminal spikes-like clusters and blooming in the month of September-October.

It occurs in grasslands, margins of field, weed in crop field in the temperate Himalaya from Kashmir to Uttaranchal at elevation 2500 m. It is propagated by seeds.

The seeds yield 1.7% total alkaloid consisting of delatine, delphaline etc. It

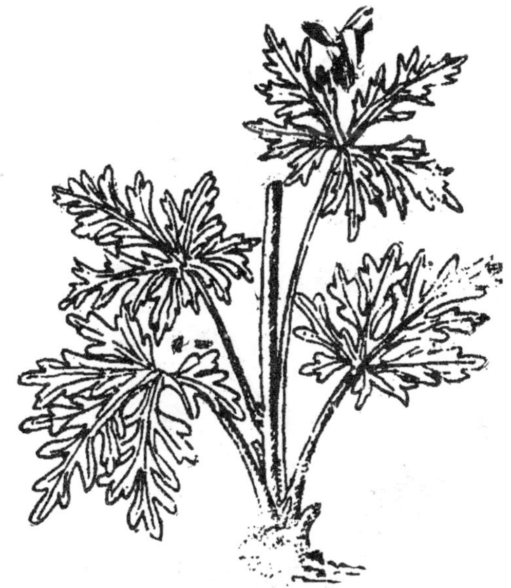

Delphinium denudatum

possesses stimulant, tonic and alternative properties and useful in toothache relief. Seeds are used as insecticides and treatment of skin eruption.

Dendrobium alpestre Royle.

Family : Orchidaceae
Trade/Common Name : Wild Orchid

An epiphytic orchid. Flowers are white.

It is found in Uttaranchal Himalaya upto 2200 m altitude. It is propogated by bulbs.

Bulbs are medicinally important. Its preparations are used to treat pimples, boils and other skin problems. The herb if also used as a tonic.

Dendrobium normale Fale.

Family : Orchidaceae
Trade/Common Name : Blue Orchid

An epiphytic orchid. Flowers pinkish white in colour.

It is found in Uttaranchal Himlaya upto 2200 m altitude. It is propogated by bulbs.

The whole herb is used as aphrodisiac and tonic.

Dendrocalamus strictus (Roxb.) Nees.

Family : Poaceae
Trade/Common Name : Male Bamboo

An arborescent unarmed bamboo.

It is found in Madhya Pradesh, Chhatisgarh, and also throughout greater parts of India. It is propagated by clums.

The clums are used as a tonic and astringent.

Dendrocalamus strictus

Derris elliptica (Wall.) Benth.

Family : Fabaceae
Trade/Common Name : Indian Tuba root

A climbing shrub with alternate leaves. Leaflets are opposite. Flowers papilionous and pod indehiscent.

It is found in West Bengal and eastern parts of India. It is propagated by seeds or root cuttings.

Roots are used as insecticides.

Desmodium gangeticum (L.) DC.

Syn. *Hedysarum gangeticum* Linn.

Family : Fabaceae
Trade/Common Name : Shalparni

A erect or ascending-prostrate undershrubs, 60-130 cm high, with somewhat angular branches, pubescent or glabrous. Leaves simple, variable, ovate-oblong or rounded. Flowers purplish or white racemes, often combined to panicles and blooming in the month of March to continue upto December.

It is found throughout the warmer parts of India. It occurs commonly in Pine forests, grassy localities, edges of fields, upto 1000 m. It is propagated by seeds or layering.

Roots are astringent and tonic, used in fever and dysentery.

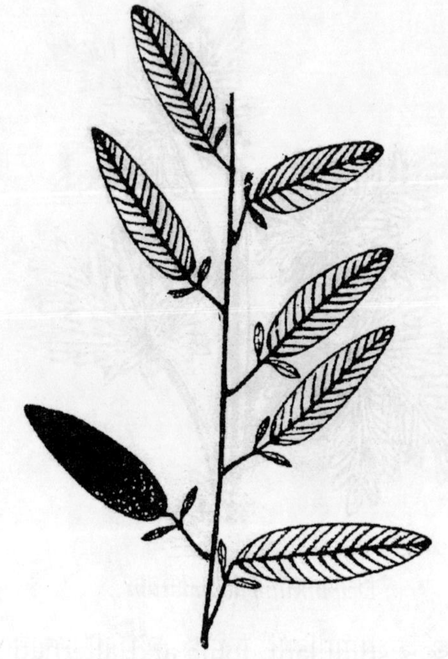

Desmodium gangeticum

Dicentra cucullaria (Linn.) Bernh.

Family : Fumeriaceae
Trade/Common Name : Duchman's Breeches.

A hardy, perennial herb with tuberous roots upto 30 cm tall. Flowers are white and sometimes yellow in colour.

It is native in Watern Himalaya. Ir is propagated by divisions of crown or cutting of tuberous roots.

Tuberous roots are medicinally important. It is used for veneral diseases, to increase urine flow or as a tonic. Locally, its poultice is applied for curing skin diseases.

Digitalis lanata Ehrh.

Family : Scrophulariaceae

Trade/Common Name : Grecian Foxglove

A biennial or perennial, erect, half hardy, unbranched herb upto 70 cm in height. Leaves broads, linear, lanceolate, mostly radical. Flowers in spikes, brownish with a white lip, showy.

It occurs in woodlands, forest clearings, grasslands, slopes. It is introduced in western Himalaya. It is cultivated as cash crop in temperate regions of Kashmir and some pockets of South Indian hills at an elevation of 1000-2500 m. It is propagted by seeds.

The leaves contain nearly 70 different glucosides. Out of these glucosides, lanatoside C and digioxin are two important glycosides used in medicine. It possesses cardiac stimulant properties. The drug promotes and stimulates the activity of all muscles. It also stimulates the digestive organ and all nervous system.

Digitalis lanata

Digitalis purpurea Linn.

Family : Scrophulariaceae

Trade/Common Name : Purple Foxglove

An erect, biennial herb upto 1.5 m tall. Leaves are ovate and hairy. The lower basal leaves are long stalked; upper leaves are almost without stalks. Flowers appear in erect, showy, usually one side spikes, white or purple in colour with ciliate lobes.

It is cultivated in Kashmir, Darjeeling and Nilgirir hills. It is propagated by seeds.

The leaves contain nearly 30 different glucosides. Out of these glucosides, lanatoside C and digioxin are two important glycosides used in medicine. The drug promotes and stimulates the activity of all muscles. It also stimulates the digestive organ and all nervous system. Locally, its infusion is used to relieve sore throat and asthma. It is administered for the treatment of congestive heart failure.

Digitalis purpurea

Dioscorea bulbifera

Dioscorea bulbifera Linn.

Syn. ***D. sativa*** **Auct. non L.**

Family : Dioscoreaceae
Trade/Common Name : Varahikand

A large, glabrous twiners, with tuberous root-stock. Stem twining to the left. Bulbils numerous, irregular, brown, warted. Leaves simple, ovate, acuminate, more or less deeply cordate at base, 9-11 nerved. Male spikes 5-10 cm long, in axillary clustered or leafless panicles. Female spikes 10-25 cm long, in axillary clusters and blooming in the month of July to continue upto September.

It is found in scrub foresta and open slopes upto 1000 m elevation. It is propagated by rhozome cuttings.

Root tubers are used for the treatment of diabetes, intestinal worms, skin diseases and as a general tonic. Sometimes is used as a substitute for 'Riddhi' and Vrddhi' of Astavarga of Ayurveda.

Dioscorea deltoidea Wall. ex Kunth.

Family : Dioscoreaceae
Trade/Common Name : Dioscorea

An extensive climber with stem twining to the left. Leaves are stalked, cordate, acute, often triangular, trigonous membranous, resembling the nose ornament (Nath) worn by ladies in the hills. Rhizomes are horizontal, borne close or often deep in the

Dioscorea deltoidea

soil (depending on the site), grayish brown in colour, with rigid scattered roots on them. Flowering occurs in the month of May to June.

It is found in forest clearings, shrubberies, slopes and rocky substrates of North-Western Himalaya from Kashmir and Punjab eastward to Nepal and China at 1500 – 3200 m. Due to over exploitation, its population has dwindled alarmingly. It is propagated by rhizome cuttings.

The rhizome contains diosgenin (4.8-8.0%) and steroid sapogenin. It is used in the treatment of various diseases particular in rheumatic diseases; ophthalmic disorder, allergic states and in idiopathic thrombocytopenic purpura. Local communities, the crushed rhizomes are given with kneaded flour and salt to oxen for general gastric complaints.

Dioscorea floribunda Mar. & Gal.

Family : Dioscoreaceae
Trade/Common Name : Bhusa

A robust climber with thick fleshy branched tuber. Flowers are small, sessile and purple in colour.

Dioscorea floribunda

It is native to Southern Mexico and Guatemala. It is commercially cultivated in Karnataka, Kerala, Tamil Nadu, Maharashtra, Meghalaya, Assam, Andaman Island and Goa. It is propagated by tubers.

Tubers is rich in glycoside saponin, yield sapogenins on hydrolysis. It contains diosgenin, damogenin, botogenin and krytogenin. Diosgenin is used for systhesis of cortisone, which is useful in the treatment of rheumatic arthritis and are in preparation of sex hormones. It is also used against asthma, nephritis, allergic manifestation and shock.

Diospyros melanoxylon Roxb.

Syn. *D. montata* Roxb.

Family : Ebenaceae
Trade/Common Name : Tendu

A shrub growing to a medium sized tree upto 20-25 m tall.Bark grayish black. Leaves elliptical, rounded at base; lower side of leaves are covered with woolly hair. Flowers dioecious, male in peduncled, 3-12 flowered, villous, simple or branched, often drooping, cymes; female solitary, axillary, subsessile, thick peduncles and blooming in the month of April to continue upto June.

It is native to India. It grows in sub-Himalayan tract in Eastern Rajasthan, Madhya Pradesh, Uttar Pradesh, Bihar, Jharkhand, Andhra Pradesh, Gujrat, Chhatisgarh and Maharashtra. It also occurs through sparingly in Tamil Nadu and Karnataka. One of the common tree of the dry mixed deciduous forests of Chhatisgarh. A characteristic tree of dry teak forests and dry sal forests in the plains. It is propagated by seeds.

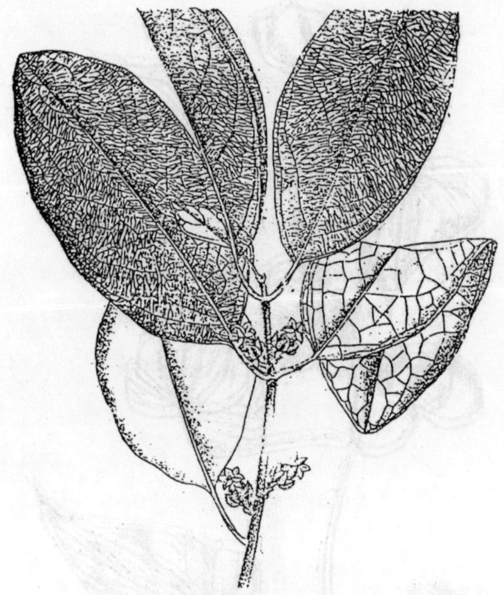

Diospyros melanoxylon

Leaves are diuretic, carminative, laxative and styptic. Dried flowers are useful in urinary, skin and blood diseases. The bark is astringent and its decoction is used in diarrhoea and dyspepsia. A dilute extract is used as astringent lotion for the eyes. Fruits are cariminative, laxative and diuretic.

Dodonaea viscosa (Jacq.) Linn.

Family : Sapindaceae
Trade/Common Name : Hop bush

An evergreen shrub; branches glandular. Leaves estipulate, simple, subsessile, 2,5-12 cm long, oblanceolate, subapiculate, base cuneate, margins entire

Dodonaea viscosa

or revolute, surface viscid-shining. Flowers small, greenish-yellow in colour borne in short terminal panicles and blooming in the month of August to continue up to November.

It is found in tropical and subtropical areas of Himalayas. It is propagated by seeds.

The leaf decoction is taken once in a day for two days to stop excess menstrual flow.

Dolichos uniflorus Lank.

Syn. *D. biflorus* Linn.

Family : Fabaceae
Trade/Common Name : Horse gram

An annual, branched, sub-erect or trailing, hairy cultivated herb with 3-foliolate leaves. Flowers appear on avery short pedicles in axils of the leaves, 1-3, yellow in colour and blooming in the month of August to continues upto September.

Dolichos uniflorus

It is cultivated in the entire sub-Himalayan tracts upto 1800 m in sunny exposed places. It is propagated by seeds.

Seeds are rich source of urea. Also contain lectins, phytoaggulutinin, globulin, coumesterol, psoralidin and considered as anthelmintic, antilithiatic, astringent, diaphoretic, diuretic, emmenogogue, expectorant, febrifuge, ophthalmic and tonic. Useful in asthma, cold, cough, colic, hiccough, effective and menstrual disorders in the form of decoction. The plants is useful in tumors, adenites, sores, burns, small pox, dysurea and rinderpest diseases.

Drosera peltata Sm.

Syn. *D. lunata* Buch-Ham.

Family : Droseraceae
Trade/Common Name : Sundew

A perennial, insectivorous, erect, delicate herb; 7-20 cm tall with small brown bulbous roots. Leaves are peltate, half moon shaped, glandular, viscid. Flowers bisexual, regular, white, in terminal, branching racemes and blooming in the month of August.

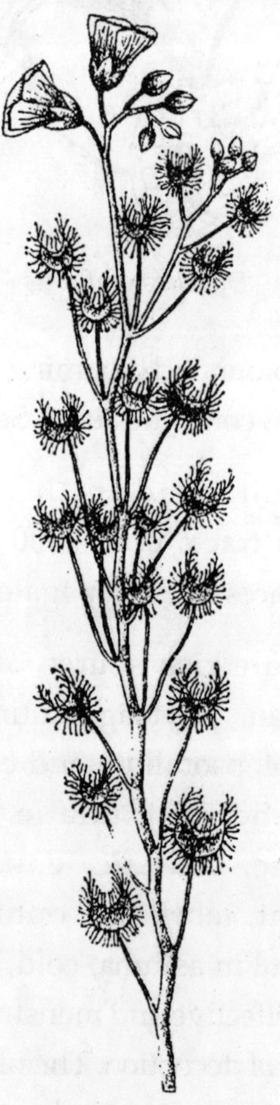

Drosera peltata

It is found in scattered form in pasture lands of temperate Himalaya between 1300-2800 m. It has no attempt so far been made to cultivate.

The herb contains resin, plumbagin and a proleolytic enzyme of the pepsin type and a yellow-brown crystalline pigment. The herb is to be used by Ayurvedic practitioners in the preparation of gold bhasma used as antisyphilitic, alternative and tonic. The air dried entire herb is used in chronic bronchitis, asthama and whooping cough, given usually as a tincture in the dose of 0.5-2.0 ml.

Drymaria cordata Willd

Family : Caryophyaceae
Trade/Common Name : Abijalo

An annual diffused or ascending herb with weak stem and dichotomously branched from the base, rooting at basal nodes. Leaves ovate, cordate, 3-5 nerved from base, entire, mucronate, stipules lacerate into filaments. Flowers small, in axillary or terminal cymes and blooming throughout the year.

It is found shady, moist slopes of sub tropical regions of Sikkim Himalaya at 1500 m elevation. It is propagated by seeds or cuttings.

The herb is used to cure sinusitis and nosal blockade. Its extract is also reported as antifebrile and laxative.

Dryopteris filix-mas Achott.

Syn. ***Aspidium filix-mas*** **Sw.**

Family : Aspidiaceae
Trade/Common Name : Male Fern

A terrestial fern upto 1.5 m tall.

It is native to Europe and found in various regions of Asia pacific. It is propagated by sores or by divisions.

Rhizomes and stipes are medicinally important. It is used as anthelmintic. Locally, its decoction is used in treating taenia and worms lodging in the alimentary canal.

Duboisia myoporoides Brown.

Duboisia myoporoides

Family : Solanaceae
Trade/Common Name : Cork-Wood Tree

A small sized tree. Leaves alternate, ovate-oblong to broad lanceolate, blunt tipped, tapering at base. Bark grey, corky, fissured. Flowers bell-shaped, rounded, white or pale lavender in colour and borne in terminal, much branched panicles.

It is natine to Australia. It is cultivated in Karnataka as commercial crop. It is propagated by seeds.

Leaves are rich source of tropane alkaloids like hyoscine, hyoscyamine, norhyoscyamine, proridine, anabasine, etc. These alkaloids are used in the modern medicine for their antispasmodic, anticholinergic, sedative, antiemetic, and mydriatic properties. It is administered as a premedication before anaesthesia and employed for sympatholytic paralysis and also used for the treatment of respiratory and intestinal disorders.

Echinops cornigerus DC.

Family : Asteraceae
Trade/Common Name : Globe-Thistle

A perennial herb. Stem leafy. Leaves deeply lobed, white cottony beneath and spiny margins. Flower-heads are arranged in spherical infloresence of numerous pale blue flower along with spine-tipped involucral bracts. Flowering occurs in the month of July-September.

It occurs grazing grounds, stony slopes, dry and srid climatic areas Ladakh Himalaya. It is propagated by root cuttings.

Echinops cornigerus

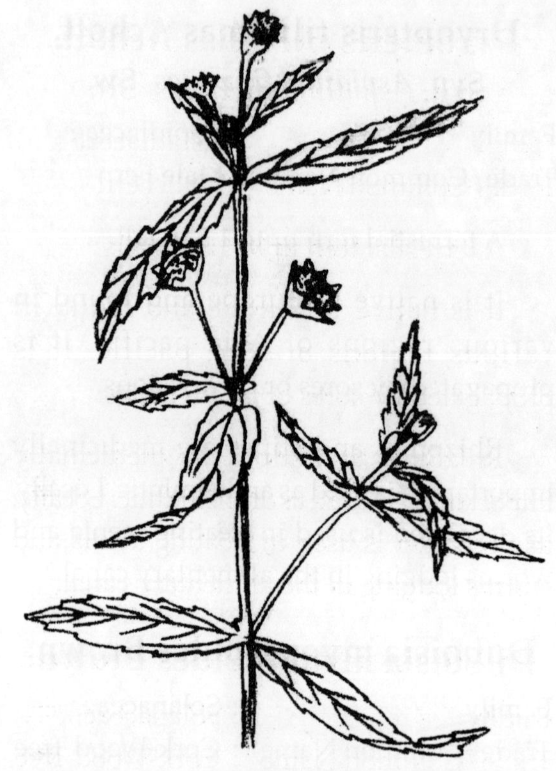

Eclipta prostrata

The roots are poisonous in nature and used in folk medicine. It is useful in skin eruptions. The root extract given to infants to promote emergence of teeth. Locally, root juice given in fever and urinary troubles.

Eclipta prostrata Roxb.

Syn. *E. alba* Hook.

Family : Asteraceae
Trade/Common Name : Bhringraj

An erect, slender, roughly pubescent herb leaves opposite, sessile, narrowly lanceolate. Heads radiate, terminal on erect stalks. Flowers white; pappus of 2-5 minute teeth ligules and blooming in the month of August to September.

It is common weed in moist situations throughout India, ascending 2000 m on the hills. It is propagated by seeds.

Leaves contain a-terthienyl-methanol, stigmasterol, b-amyrin, mixture of desmethyl wedelolacetone. It is used as tonic, deobstruent in hepatic and spleen enlargement. The plant juice in combination with aromatics is administered for jaundice. Leaves are useful in scorpion sting. Root is emetic, purgentive; applied externally as antiseptic to ulcers and wounds. The oil prepared with this herb is of great repute as a hair dye and has cooling affect on the brain.

Elettaria cardamomum (L.) Maton.

Family : Zingiberaceae
Trade/Common Name : Cardamom

An erect, perennial herb upto 3 m tall. Leaves are large, narrow, with strong medium nerves. Flowers are white or pale green in colour and appear from near the base of the stem and lies on the ground.

It is native to India and Ceylon. It occurs wild in moist forest and hilly regions of Karnataka and Kerala. It is commercially cultivated in Kerala, Karnataka and Tamil Nadu. It is propagated by seeds.

Elettaria cardamomum

Seed yield an essential oil which constitute eucalyptol, sabinine, d-terineol. It promotes digestion and is administered with purgatives and as a flavouring agent. The tincture of cardamom is administered to relieve winds from stomach. It is reported to be a cardiac stimulant, and checks nausea and vomiting.

Embelia ribes Burm.f.

Family : Myrsinaceae
Trade/Common name : Embelia, Baibidang

A large extensive traggler with elongate branchlets hanging over. Leaves alternate, elliptic or elliptic-lanceolate. Flowers white in terminal and axillary panicle and blooming in the month of March-July.

It is found in sub-tropical regions of Himalaya but rarely occurs in hilly parts of South India. It is propagated by seeds.

Embelia ribes

Roots infusion is given for coughs and diarrhoea, decoction of dried fruits is used for fevers and diseases of chest and skin. Dried fruit is anthelmintic, astringent, tonic; used in scorpion sting and snake bite. Leaves are useful in pruritus, skin diseases and leprosy.

Embelia tsjeriam-cottam DC.

Family : Myrsinaceae
Trade/Common Name : Baya-birang

A small tree with scandent branches and dark-brown bark. Leaves are narrow at both ends upto 12 cm long, lower surface reddish and hairy on nerves. Flowers are pale green in colour, minute in size and borne in axillary or extra axillary, pubescent racends. Fruits are small, globose, pointed and red when ripe.

It is found in Eastern India and Deccan Peninsula. It is propagated by seeds.

The dried fruits has medicinally important. It possess antibacterial and antitubercular properties. It is laxative and effective in the treatment of tapeworms. Sometimes adulterated with black-pepper.

Emblica officianalis Gaertn.

Syn. ***Phyllanthus emblica*** **Linn.**

Family : Euphorbaceae
Trade/Common Name : Emblic Myrobalan

A deciduous tree nearly 12 m tall. Leaves small narrowly linear and resemble pinnate leaves. Flowers are fleshy, globose and pale yellow and blooming in month of March to continue upto May.

Emblica officianalis

It is distributed throughout tropical and subtropical India, chiefly in the dry deciduous forest ascending upto 1350 m in the Himalaya. It is propagated by seeds.

It is a rich source of vitamin C. Seeds contain fixed oil, phosphatides and an essential oil. Fruits and leaves contain tannins, polyphenolic compounds, corialgin and ellagic acid. It is diuretic, laxative, cardiac, astringent and liver tonic. Useful in anaemia, diarrhoea and dysentery, dyspepsia, haemorrhage, inflammation of the eyes, irritability of the blodder, jaundice, leucorrhoea, menorrhagia and discharge of the blood from the uterus; affects blackening

Cichorium intybus

Cinnamomum camphora

Cinnamomum tamala

Cissampelos pareira

Clerodendron serratum

Colchicum luteum

Coleus forskholii

Commiphora mukul

Coriandrum sativum

Corydalis govaniana

Costus speciosus

Crocus sativas

Curculigo orchioides

Curcuma domestica

Cuscuta reflexa

Cymbopogon flexuosus

Cymbopogon winterianus

Cynodon dactylon

Cyperus rotundus

Dactylorhiza hatagirea

Daucus carota

Dioscorea bulbifera

Dendrocalamus strictus

Drymarla cordata

Echinops cornigerus

Eclipta prostrata

Emblica officinalis

Ephedra gerardiana

Eugenia caryophyllata

of hair. It is a chief ingredient of Chyavanprash Avleha, a tonic and restorative in Indian medicine.

Ephedra garardiana Wall. ex Stapf

Syn. ***E. vulgaris*** **Hook. f. (non A. Rich.)**

Family : Ephedraceae
Trade/Common Name : Somlata

A slow growing rigid, tufting shrub, 30-60 cm, often having variable size with numerous, densely clustered, erect, slender, smooth green, jointed branches arising from a branched woody base in whorls. Internodes of branchlets are 1-4 cm long and 1.2 mm in diameter. Branches have scales at the joints. Male cone ovate, solitary, 4-8 flowers. Female cone usually solitary and blooming in the month of May to continue June.

Ephedra garardiana

It occurs in stony slopes, gravel terraces, rocky substrates and drier regions of temperate and alpine Himalayas, from Kashmir to Sikkim at 2800-5100 m altitude. It is propagated by seeds, suckers and layers.

The aerial shoots/twigs contain ephedrine and pseudo-ephedrine. Total alkaloid content in Indian ephedras varies from 0.28-2.79%. The alkaloid content increases with the age of plant and the best period to collect the green twigs is, when the plants are 4 years old and are in blossom. The dried twigs which from the official drug are used in asthma, hay fever and rashes of allergic origin. Liquid extracts of the plant is used for controlling asthmatic paroxysms.

Epipactis latifolia Wall.

Family : Orchidaceae
Trade/Common Name : Helloborine

A terestial orchid with stout stem. Leaves are broadly elliptic to long. Inflorescence are rather dense sometimes twisted, few to many flowered with large floral bracts. Flowers are greenish variously marked with purple, white, bell shaped.

It is native to North America, Asia and Europe. It occurs Kashmir and Sikkim between 2000 to 3500 m altitude. It is propagated by seeds or cuttings.

Rhizome are useful. It is used to treat nervous disorders. Infusion of leaves gives in intermittent fever. Rhizome is regarded as an aphrodisiac.

Equisetum debile Roxb.

Family : Equisetaceae
Trade/Common Name : Fox-tailed Asperagus

An aquatic, green herb. Leaves narrow, rush like.

It occurs in temperate regions of Uttaranchal upto 2100 m elevation. It is propagated by division of root cutting.

It posesses cooling and diuretic properties and it is used to treat gonorrhoea.

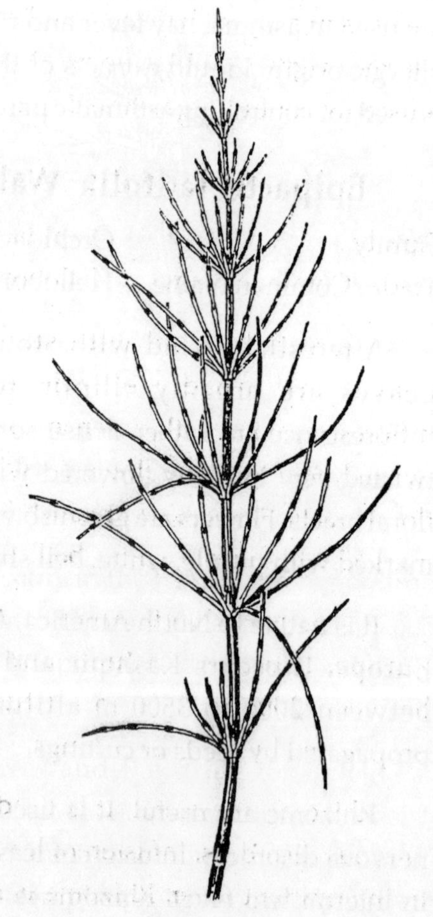

Equisetum debile

Eria musicola Linn.

Family : Orchidaceae
Trade/Common Name : Wild Red Orchid

A small epiphytic herb with globose and crowed pseudobulbs. Leaves are only 2-3, petioled, oblanceolate upto 3 cm long. Flowers are very small, yellowish green arise on raceme.

It is found in Jharkhand and Khasia hills between 1000-1300 m altitude. It is propagated by seeds.

Pseudobulbs are used in medicine. It is used to treat chest, heart, lung, eye, ear and mental problems.

Erigeron canadensis Linn.

Family : Asteraceae
Trade/Common Name : Beach Aster

An annual, erect, much branched herbs, 20-120 cm high; stem ribbed, patent-hairy. Leaves sessile or shortly petiolate, linear-lanceolate, acute, apiculate entire, glabrous or loosely hairy. Heads 5-6 mm across, peduncles, yellowish, in branched panicles. Ray florets pale or rosy-purplish, hardly exceeding the pappus. Disc florets tubular and blooming in the month of February to continue up to September.

It is dispensed over Upper Gangetic Plains, Western Himalaya, Punjab, Assam and Western Ghats. It is propagated by seeds or by divisions of root.

Herbaceous part of the plant yield 0.25% on stream distillation. The oil contains d-limonene, terpinyl acetate and citral. It is homeostatic and used rarely in some medicinal preparations.

Erodium cicutarium (L.) L'Herit.

Family : Geraniaceae
Trade/Common Name : Common Stork's Bill

A hardy biennial hairy herb with compound leaves. Leaflets are without any stem. Flowers are purple in colour.

It is native to Mediterranean regions. It is propagated by divisions of roots.

Mature leaves and blooming plants are useful. A tea prepared from leaves are used to increase perspiration and urine flow, to treat uterine, haemorrhage and water retention. The plant is also effective to treat pains of rheumatism.

Erythrina indica Lam.

Syn. *E. variegata* Linn

Family : Palilionaceae
Trade/Common Name : Coral Tree

A fast growing, densely branched, deciduous tree up to 15 m height. Bark is yellowish-grey and smooth. Stem is erect with black spines. Leaves are trifoliate; leaflets are broadly ovate, acute and smooth. Flowers are pea shaped, red in colour and appear during February to May in dense racemes.

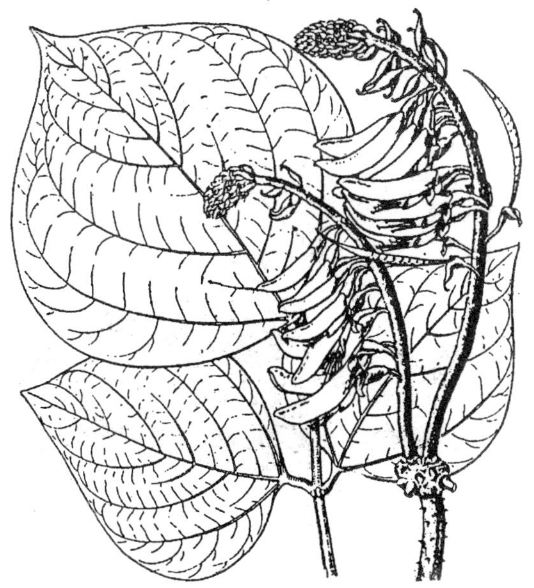

Erythrina indica

It is found in the forest of Andaman and Nicobar Islands. It is propagated by stem cuttings and by seeds.

Coral tree is used for the some pharmaceutical preparations.

Eucalyptus citriodora Linn.

Family : Myrtaceae
Trade/Common Name : Eucalyptus oil

A large tree, often attaining a great height with a smooth whitist to pale pink bark.

It is native of Queensland and now grown in most of the hills stations of South India. It is propagated by seeds. It can be identified by its characteristic fruit and lemon scented leaves.

Leaves yield an essential oil (0.75-1.25%). The oil contains cineole (75-85%). It

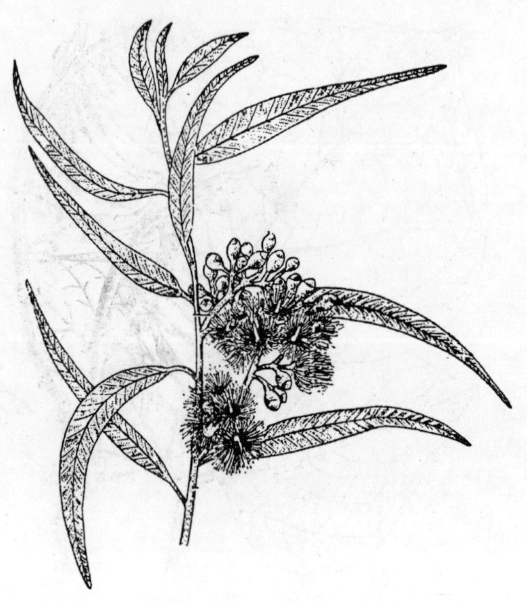

Eucalyptus citriodora

Eucalyptus globules

is widely used in soap, perfumery and cosmetic industries and for the isolation of 'hydroxy citronellal' used in the manufacture of high grade perfumes.

Eucalyptus globules Labill.

Family : Myrtaceae
Trade/Common Name : Blue Gum

A erect, tall tree. Bark is grayish or bluish white in colour. Leaves are up to 20 cm long, pendulous, lanceolate, cordate-ovate, stem claspine and glaucous. Flowers are bluish-white in colour, either solitary or 2 or 3 in number.

It grows well in Himachal Pradesh, Kumaon hills, Darjeeling, Shillong and South Indian Hills. It is propagated by seeds or cuttings.

Leaves and terminal branches yield an essential oil (0.75-1.25%). The oil is colourless or pale yellow. Main constituent of the oil is cineole. It possess deodorant and antiseptic properties. It is used for medicinal purposes for the treatment of lung diseases.

Eugenia caryophyllus Spregal.

Family : Myrtaceae
Trade/Common Name : Clove

An evergreen tree with straight trunk upot 10 m high. The flower buds have a slightly cylindrical base and is surrounded by plump ball like upopened corolla which is surrounded by the four toothed calyx. The flowers are fleshy purplish in colour.

It is native to India, Sri Lanka, Madagascar, East Africa, and Indonesia. It is cultivated in Kerala and Tamil Nadu. It is propagated by seeds.

Eugenia caryophyllus

narrower than sepals and the tip is as long as sepal.

It is found in submontane to montane Himalaya, from Kashmir to Sikkim and adjacent plains. It is propagated by divisions of roots.

Tubers are medicinally important. It is aphrodisiac and used to treat pulmonary and cardiac problems. It is also useful in cough and cold.

The leaves, unopened buds and unripe fruits yield an essential oil. The clove buds contain volatie oil (15%) and protein, carbohydrate, minerals and vitamins A, B and C. The essential oil of dried flower buds contain eugenol (90%) and also eugenol acetate, caryophyllene, acetyleugenol, vanillin and furfural. It is used as culinary spice. It has expectorant and antiemetic properties.It was also used as anodyne and antiseptic in dentistry. It is carminative and stimulant and used in flatulence and dyspepsia. The clove oil possess antiseptic and anti-bacterial properties.

Eulophia campestris Wall.

Family : Orchidaceae

Trade/Common Name : Sudhamuli

A terrestial orchid with deformed pseudoblubs. Leaves linear, long. Flowers are pale with darker lines, petals are

Eulophia campestris

Eulophia herbacea Lind.

Family : Orchidaceae
Trade/Common Name : Dudhiya

A terrestial orchid with tuberous roots. Stem long, slender and sheathed. Leaves are linear-lenceolate, many nerved upto 30 cm long and 8 cm broad. Flowers are long, white with purple nerves.

It is found on the grassy slopes, edges of scrub jungles in Uttaranchal, West Bengal, Orissa and Karnataka. It is propagated by tubers.

It is used in preparation of Ayurvedic medicines and used as general tonic.

Eupatorium cannabinum Linn.

Family : Asteraceae
Trade/Common Name : Mugwort

A undershrub, 1-2 m high; stem subterete, raddish-brown, clothed with glandular-hairs. Leaves broadly ovate cuneate, glandular-hairy crenate-serrate, base cuneate; petioles 2-3 cm long. Heads discoid, 5-7 m across, several in pedunculate, terminal, viscid-hairy corymbs. Florets white or dull-white, 2-3 mm long. Achenes black, oblong, 5-angled, glabrous. Pappus hairs 10, dirty-white and blooming in the month of Feburary to continue up to August.

It is found as weed in open sunny slopes of sub tropical regions of Sikkim Himalaya around 1600 m elevation. It is rarely found in inside the dense forest. It is propagated by runners.

Eupatorium cannabinum

Locally, the leaves and tender stems are crushed fresh and their juice is applied to cut and bruises. Sometimes when the wound is large the squeezed remains of the plant is placed over the wound and poultice applied. The bleeding stops immediately and the wound is protected from infection.

Euphorbia helioscopia Linn.

Family : Euphorbiaceae
Trade/Common Name : Sun spurge

An annual herb upto 40 cm in height. Leaves obovate to spathulate. Latex white. Glands of involucre yellow. Flower-heads yellow in flattopped cluster.

Euphorbia helioscopia

It occurs waysides, wastelands, orchards, as weed in crop fields of temperate Himalaya. It is propagated by seeds.

The herb is poisonous in nature and having anthelmintic properties. Locally, the seeds given with roasted pepper in cholera. The latex is applied on wounds and sprains as antiseptic and applied on warts.

Euphorbia hirta Linn.

Family : Euphorbiaceae
Trade/Common Name : Dudhi

An erect herb upto 50 cm tall with greenish yellow or white flower. Stems covered with yellowish hairs. Leaves are small in opposite pairs.

Euphorbia hirta

It is found throughout India and common in warmer regions. It is propagated by seeds.

The herb is used to treat colic trobles, dysentery, cough, asthma, worms and vomiting. It possess vermifuge properties.

Euphorbia parviflora Linn.

Family : Euphorbiaceae
Trade/Common Name : Round Milky Plant

An annual herb up to 90 cm high. Stem simple or branched. Leaves alternate, sessile, ovate or oblong-lanceolate, acute or obtuse, base narrowed, membranous. Involucres in terminal or axillary peduncled cymes and blooming in the month of May to continue up to August.

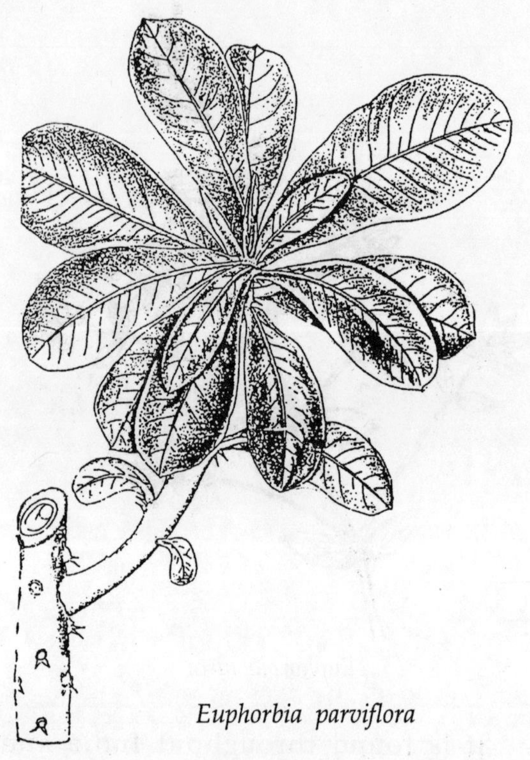

Euphorbia parviflora

It is native to India. It is propagated by seeds.

The dried leaf infusion is used to treat diarrhoea, dysentery and leucorrhoea.

Euphorbia pilosa Linn.

Family : Euphorbiaceae
Trade/Common Name : Choplas

A perennial, glabrous herb, 15-40 cm high; stem several from the base, with thick rootstock. Leaves alternate, sessile, linear-lanceolate or oblong, ontise, retuse or mucronate, subumbellate cyme. Involucres shorty pedicelled, glabrous, 4-5 mm across; lobes 5; glands yellowish crescent-shaped,

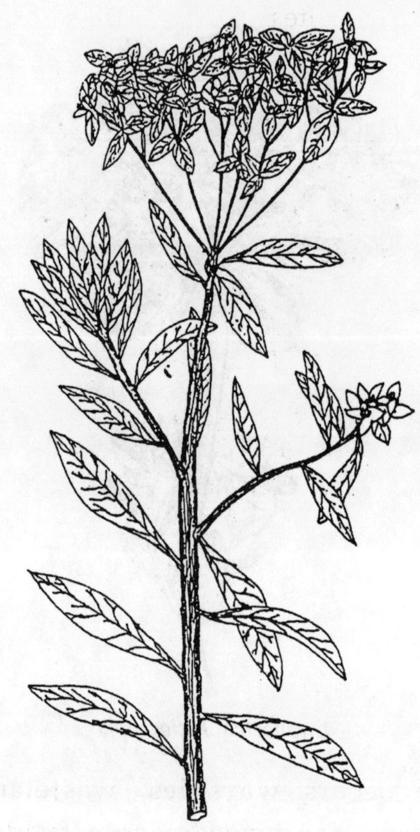

Euphorbia pilosa

2-horned and blooming in the month of April to continue up to May.

It is native to Europe and North Asia. It is found in moist temperate zones of Himalaya. It is propagated by seeds.

The roots are purgative and vermifuge and used to treat fistular sores.

Euphorbia royleana Boiss.

Family : Euphorbiaceae
Trade/Common Name : Sulu

An shrub with 5-7 angled branches, spinous. Flowers are yellow in colour.

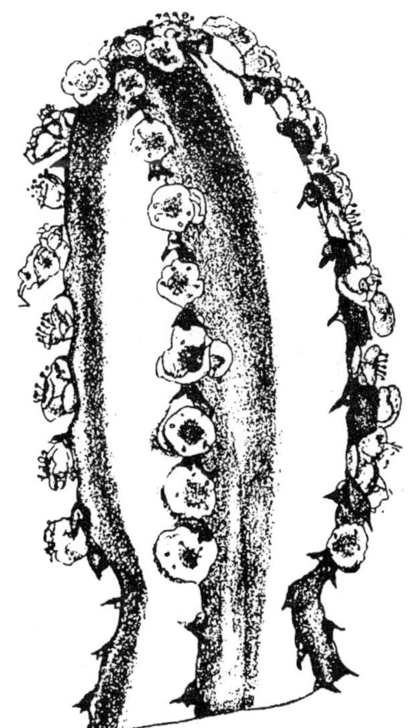

Euphorbia royleana

It occurs in Uttaranchal between 500-1500 m in the rocky and exposed valleys.

The latex is filled in the hollow cavities of decayed teeth to check the infection. The latex is cathartic and anthelmintic.

Euphrasia officinalis Linn.

Family : Scrophulariaceae

Trade/Common Name : Eyebright

A small branched herb upto 45 cm in height. Leaves ovate, saw-toothed. Stem with slender branches from the base and terminal spike-like clusters of white and purple tinged 2-lipped flowers with yellow throats.

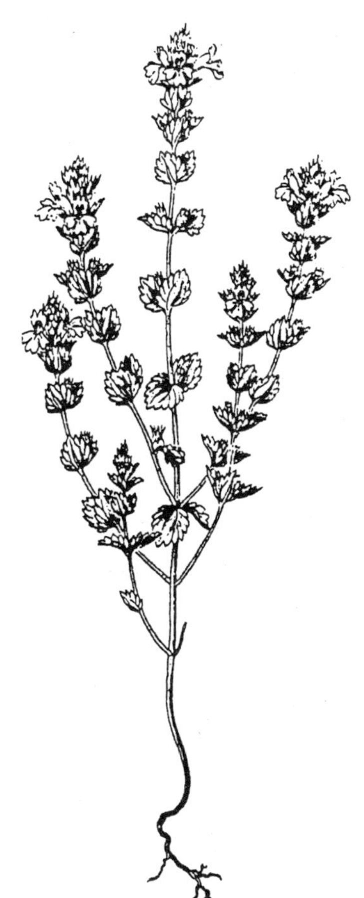

Euphrasia officinalis

It occurs damp and shady sites in forests, waysides from Kashmir to Uttaranchal in tempertae Himalaya at elevation of 3000 m. It is propagated by seeds.

The dry herb contains inroganic substance (water soluble, 20%), volatile oil (0.07%), tannins (12.4%), vitamin C, an aromatic resinous substances and a glycoside. It possesses bitter, astringent and tonic properties. It is useful conjuctivities and jaundice.

Fagopyrum aibotrys (D. Don) Hara.

Family : Polygonaceae
Trade/Common Name : Fafrya

An erect, hardy, annual herb.

It is native to Asia. It is found in poach of Uttara Kashi district of Uttaranchal between 2000 to 2700 m altitude. It is propagated by seeds.

Ethno-medicinally, the paste of crushed leaves is applied on the insect bite. To inhale the smoke of seeds is believed to be good for headache.

Fagopyrum esculentum

Fagopyrum esculentum Moench.

Family : Polygonaceae
Trade/Common Name : Buckwheat

An annual purplish herb 1-1.5 m, tall, with alternate, hastate, acute leaves; flowers in axillary or terminal cymes pinkish white, fragrant, dimorphic and self sterile and blooming in the month of July-September.

It occurs throughout hilly regions and also cultivated as a rainy season crop in mid and high altitude upto 2600 m. It is propagated by seeds.

Leaves contain 3.4-4.7%, while fruits yield about 1.3% rutin, fagopyrin, flavanol, fagomine, alanine and various flavanone glycosides. It is used in modern medicines as antihaemorrhagic due to presence of glucoside rutin. Decoction of root is used in rheumatic pains, lung diseases and typhoid; while the juice is useful in urinary disorders. Rutin is valued as a cure for capillary fragility of the heart.

Ferula asafoetida Vatke.

Family : Apiaceae
Trade/Common Name : Hing

A perennial herb upto 1.5 m tall with a stout tuberous and fibrous rootstock. Leaves long upto 10 cm linear segments, finely toothed. Flowers are yellow in colour.

It is native to Afghanistan, Iran and Kashmir. It occurs open slopes, rocky substrate of temperate Himalaya. It is propagated by root cutting.

Hing is extracted from tap root of 4-5 years old plant. The root contains resin (69.98%), gum (9.21%) and essential oil (14.8%). The essential oil contains camphene

Ferula asafoetida

It is native to Afghanistan, Iran, Tibet and Kashmir. It occurs open slopes, rocky substrate of temperate Himalaya. It is propagated by seeds or cuttings.

Oleo-gum resin from rhizomes and roots are medicinally important. It contains gum (25%) and also volatile oil and resin.

It is stimulant, antispamodiac, carminative, expectorant and laxative. It is used in indigestion, whooping cough, flatulent colic, infertile convulsions etc. It is stimulant for nervous and respiratory system and prescribed in bronchitis and pneumonia in children.

(25%), d-pinene (60-90%), azulene (5%). The gum is applied to wounds and bruises. It stimulate intestinal and respiratory tracts and nervous system. It is also used to treat chronic bronchitis, asthma, whooping cough and in given as enema for intestinal flatulence. It is also employed in hysterical and epileptic affections, in cholera and as sedative in heart diseases.

Ferula narthex Boiss.

Family : Apiaceae

Trade/Common Name : Hing

A perennial herb with robust, carrot-shaped roots. Lower leaves are simple upto 60 cm long, upper divided into segments and young leaves are densely hairy. Flowers appear in large terminal clusters, small and yellow in colour.

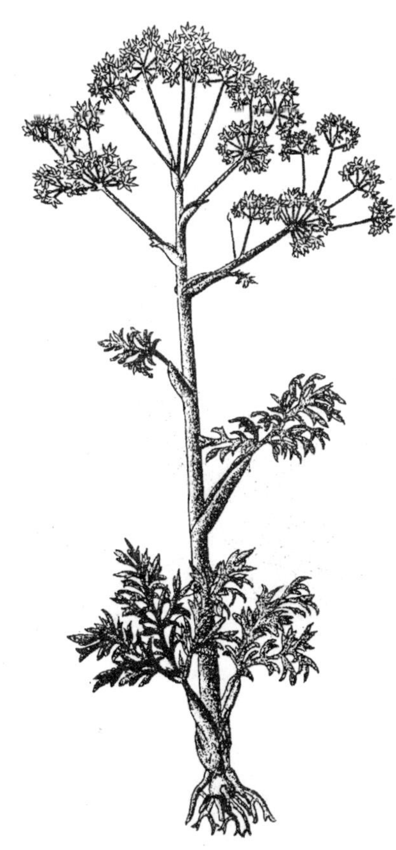

Ferula narthex

Ficus grandiflora Ehrh.

Family : Fagaceae
Trade/Common Name : Beech

A tall, hardy, deciduous tree upto 35 m height. Leaves are silky when immature, bluish green above upto 12 cm long.

It is found throughout India. It is propagated by seeds.

Mature leaves and stem bark is medicinally important. A concoction of leaves is applied to burns, scalds and frost bite. Fresh bark preparation is used to cure rashes.

Ficus grandiflora

Ficus sylvatica Linn.

Family : Fagaceae
Trade/Common Name : Common Beech

A perennial deciduous tree upto 25 cm tall.

It is native to Europe. It is found as homogeneous forest in Kulu valley, Nilgiris and elsewhere for ornaments. It is propagated by seeds.

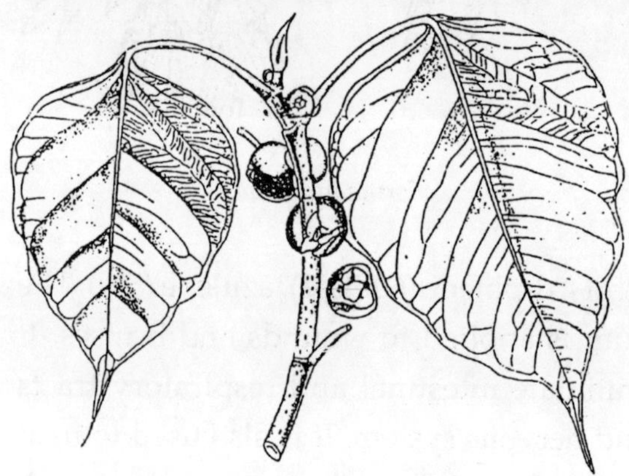

Ficus sylvatica

Beech creosote is used in medicine for treating tooth decay. The wood is used in making artistic material.

Foeniculum vulgare Mill.

Family : Apiaceae
Trade/Common Name : Fennel

A biennial to perennial smelling, stout herb upto 2 m tall. Leaves green or yellow green, finely divided segments, aromatic. Flowers yellow in umbels and blooming in the month of January to continue upto March.

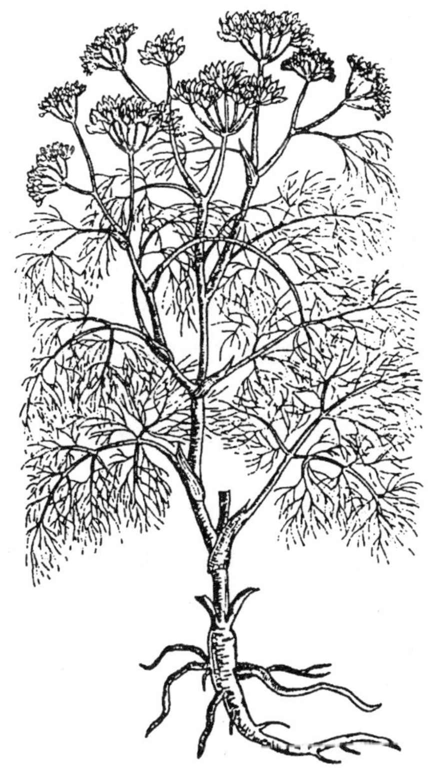

Foeniculum vulgare

It occurs wastelands, dumpyards, roadsides, orchards and vineyards in temperate Himalaya as a weeds. It is also cultivated in field as spice. It is propagated by seeds.

Seeds yield an essential oil (2.5-6.5%) which contain d-piene, camphene, d-L-phellandrene, dipentine etc. Seeds possesses stimulant, stomachic, carminative and purgative properties. It is a folk remedy for aerophagia. Leaves diuretic. Oil vermicide, used to flavour many food items. The juice made from roots is febrifuge, purgative, sudorific and carminative. It cures asthma and windy colic and useful in diseases of chest, spleen and kidney.

Fragaria virginiana Duch.

Family	: Rosaceae
Trade/Common Name	: Virginia Strawberry

A low growing perennial herb. Leaflets are three. Flowers are with fine white petals. Fruit is a small, light scarlet in colour.

It is native of America. It is found in various parts of India. It is propagated by seeds or division.

Leaves, fruits and roots are useful. Tea prepared from leaves is used for diarrhoea. Fruits are used to treat kidney stones and gout. Roots are used to increase urine flow and to treat gonorrhoea.

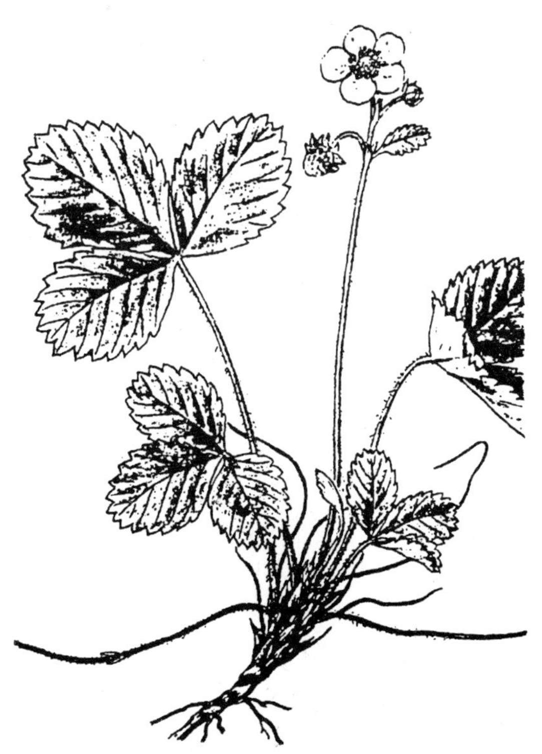

Fragaria virginiana

Fritillaria roylei Hook.

Family : Liliaceae
Trade/Common Name : Kashirkakoli

A perennial herb upto 60 cm in height. Bulbs small. Leaves linear-lanceolate often long-pointed, opposite or 3-6 in a whorl. Flowers yellowish-green to brownish purple, pendulous, 2-4 or rarely solitary and blooming in the month of June-July.

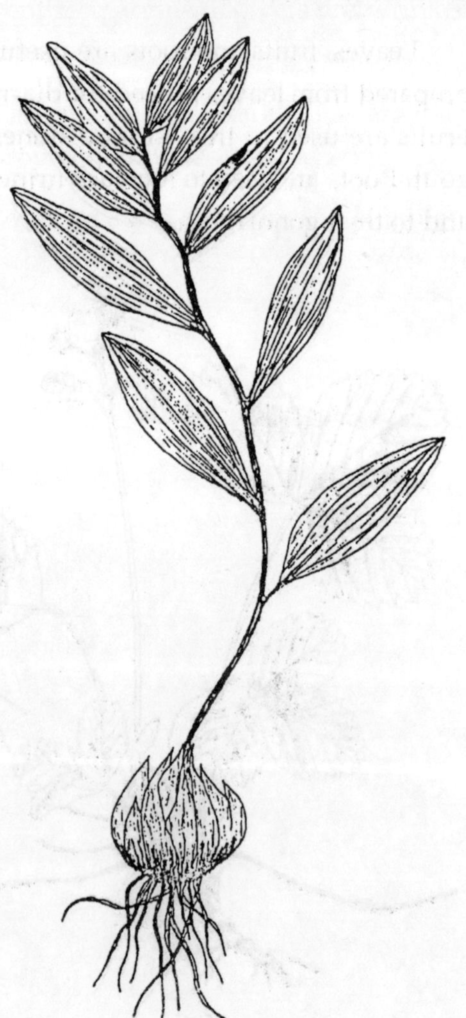

Fritillaria roylei

It occurs in alpine slopes, plateaus, grasslands, shrubberies from Kashmir to Uttarachal (Western temperate Himalaya). It is propagated by bulbs.

The bulb contains peimine, peiminine, peimisine, peimiphine, peimitidine and a natural propeimin and a sterol. It possesses antipyretic, expectorant and lactagogue properties. It is useful in dried orange skins for asthma, tuberculosis.

Fumaria parviflora Lam.

Syn. *F. indica* (Haussk.) Pugsley: *F. officinalis* Linn.

Family : Fumariaceae
Trade/Common Name : Fumitory

A pale green, diffuse, much branched annual herb. Leaves much divided,

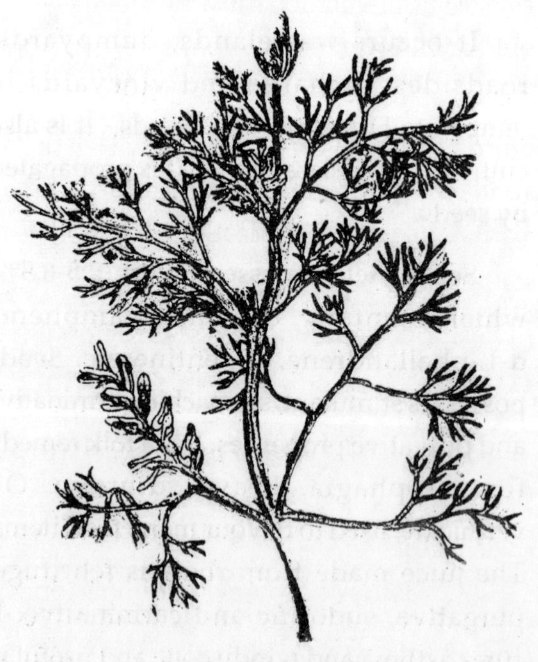

Fumaria parviflora

segments very narrow, flat. Flowers are small, purplish, in lax flowered recemes and blooming in the month of February to continue March.

It occurs common over the greater parts of India, mostly as weeds of orchards, cornfield, waste heaps and wheat field growing in the winter season. It is propagated by seeds.

The whole plant contains major alkaloid protopine along with phenolic and non-phenolic alkaloids including cryptopine, sanguinarine, fumaridine, fumaramine, parfumidine etc. The dried plant is anthelmintic, diuretic, and diaphoretic in low fever and is used to purify blood in skin diseases. In Unani system it is reported to be used in liver trouble. Extract of whole herb in water is recommended in small doses as cooling *sherbat* in fevers.

Gaultheria fragrantissima Wallich.

Family : Ericaceae
Trade/Common Name : Winter Green

A much branched evergreen shrub of 3 m tall. Leaves are leathery, broad, toothed margins, dotted with glands. Flowers are greenish white, small and appear in axillary branches.

It is found in sub-tropical to temperate zones of India between 1500 to 2500 m altitude. It is propagated by seeds or by removal of rooted offsets.

Gaultheria fragrantissima

Leaves yield an volatile oil which contitute 90-95 methyl salicylate. It contain paraffin, aldehyde, ketone, ester and secondary alcohol. This oil is added to certain ointment, are liniment to counteract their irritating effects. It is stimulant and carminative properties, and used to treat hookworms.

Gelsemium sempervirens (L.) Ait.f.

Family : Logaminaceae
Trade/Common Name : False Jasmine

A much branched, perennial, evergreen woody climber. Leaves are aromatic, lence shaped and short stalked. Flowers are tubular, attractive yellow and fragrant. The peduncle bears numerous bracteoles. Rhizomes are cyclindrical with peculiar odour, brown or yellowish orange in colour.

It is native to America and Maxico. It grows various parts of India. It is propagated by seeds or cuttings.

Flowers, roots and rhizomes are medicinally important. Roots and rhizomes preparations are used to treat fever, pain and also used as a adepressant of certain nervous system. A tea prepared from frowers is said to cure coughs, stomach pains, shortness of breath, pleurisy and also help in child birth. It is also used to treat migraine.

Gentiana kurroo Royle.

Family : Gentianaceae
Trade/Common Name : Indian Gentian

A small, decumbent, tufted herb with perennial root stock, stout, giving out numerous arching stems, 5-30 cm. Basal long; stem leaves 2-3 pairs, narrow, linear about 2 cm long. Flowers handsome, blue often spotted with white and blooming in the month of August to continue October.

It occurs temperate to alpine grassy slope, plateaus, open placess from Pakistan to Uttaranchal at the altitude 1800-2700 m. In Uttaranchal, it has become sporadic in sub-alpine to alpine meadows. It is propagated by seeds, cutting or by root division.

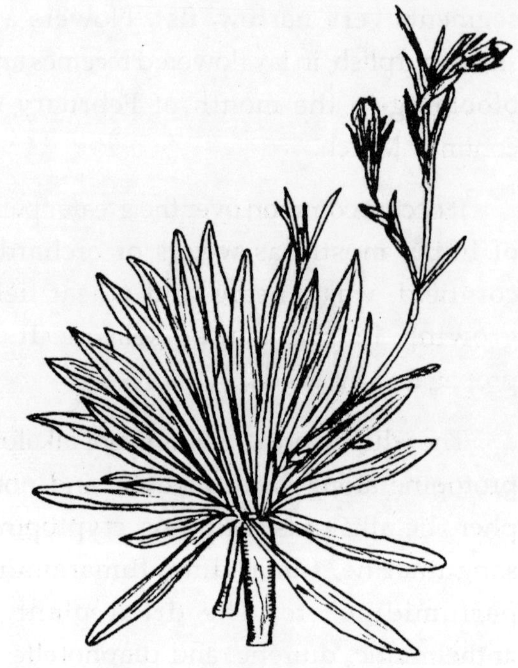

Gentiana kurroo

The root contains gentiopicrin, gentianic acid, a pectin and a sugar. It is used as a bitter tonic, stomachic and febrifuge and also considered to be an anthelmintic, blood purifier, carminative, diphoretic, digestive, useful in hepatic disorders and urinary affections.

Gentiana lutea Linn.

Family : Gentianaceae
Trade/Common Name : Yellow Gentian

A hardy perennial herb upto 1 m tall. Flowers are yellow in colour.

It is native in the Southern and Central Europe. It grows in North-Western Himalayan region of India at an altitude of 2500-3500 m. It is propagated by seeds, cutting or by root division.

Gentiana lutea

Dried rhizomes constitute the drug. It contains glucosides likes gentiopicrine; also gentiomarine and anarogentine. It is used as antimalaarial. Bitter tonic is prepared from roots, is used in dyspepsia and anorexia.

Gentiana villosa Linn.

Family : Gentianaceae
Trade/Common Name : Sampson's Snake Root

A perennial plant with ascending stems and lance-shaped leaves. Tubular shaped, greenish white to purplish-green flowers appear in the top of the plant in cluster.

It is native to America. It is found in hilly regions of Indian Himalayas. It is propagated by seeds or divisions.

Roots and rhizomes are used to treat indigestion, rheumatism and also as tonic for nervous distress.

Geranium nepalense Sweet

Family : Geraniaceae
Trade/Common Name : Ratanjot

A perennial, pubescent, diffuse herb, 15-45 cm long, prostrate, often ascending and branches rooting at the joints. Leaves orbicular 3.5-7.5 cm across, palmately 3-5 lobed. Flowers purplish red and blooming in the month of May to continue August.

It occurs frequently in temperate forests and pasture lands in Himalayan region

Geranium nepalense

between 1500-3000 m. It is propagated by seeds and root slips.

The root contains red colouring matter, tannins, gallic acid, quercetin and citiric acid. It is used as an astringent and also used in the diseases of kidneys.

Geranium ocellatum Cambess.

Family : Geraniaceae
Trade/Common Name : Kaphlya

A prostrate, annual, diffused-ascending, hairy herb with reddish purple branches, 5-20 cm high. Leaves crowded, long petioled; upper ones in pairs, short petioled; orbicular, palmately 5-7 lobed, segments, 3-lobed, toothed, sparsely pubescent; petioles; stipules ovate. Flowers purplish-white with a dark violet-purple center, solitary or in pairs; pedicles, lanceolate, acute and blooming in the month of February-April.

It is found in moist places in the Himalayan regions upto 2000 m altitude. It is propagated by seeds, stem cuttings and division of roots.

The herb has astringent and diuretic properties. The roots supposed to be antiseptic; extract given in liver trouble and fever.

Geranium walliachinum D. Don. ex Sw.

Family : Geraniaceae
Trade/Common Name : Laljari

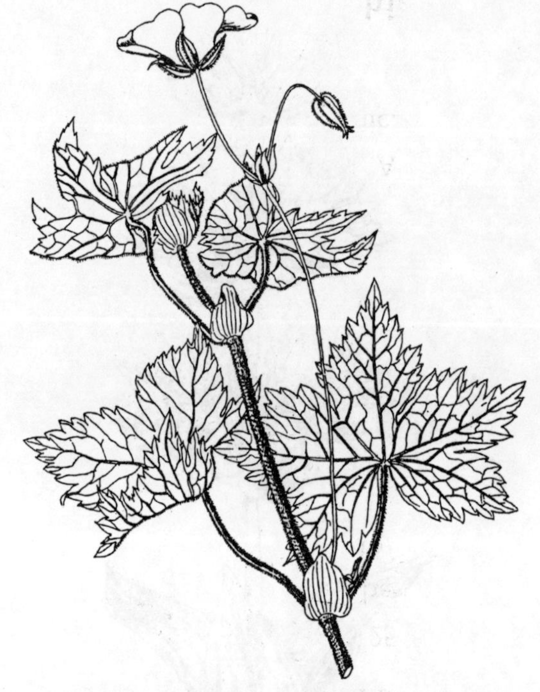

Geranium walliachinum

A perennial herb with hairy outgrowth upto 1.5 m height. Flowers are blue in colour.

It is found in temperate regions of Uttaranchal between 2200-3400 m altitude. It is propagated by seeds, cutting of roots.

Roots are medicinally important. It is given in the stomach disorders of infants and useful in cut bruises. Root juice in gonorrhoea and ophthalmia.

Gerbera lanuginosa (DC.) Schu.

Family : Asteraceae
Trade/Common Name : Transvaal Daisy

A small perennial herb. Leaves are speading from the base. Flowers are white and red in colour.

It is found in sub-tropical hills of Uttaranchal Himalayan region upto 1800 m elevation. It is propagated by seeds.

Leaves are used as tincture and for staunching wounds. Its paste plastered on bone fracture.

Geum elatum Wall.

Family : Rosaceae
Trade/Common Name : Avens

A hardy, perennial tufted herb upto 45 cm tall with fiberous, woody root stock. Leaves are radical. Flowers are bright yellow on corybose cymes.

It occurs abundantly in alpine and sub-alpine pastures and open slopes of Himachal Pradesh between 3000-4200 m elevation. It is propagated by seeds or root cuttings.

Roots are medicinally important.

Geum elatum

Geum urbanum Linn.

Family : Rosaceae
Trade/Common Name : Clove Wort

A hardy perennial herb. Flowers are pale yellow in colour.

It is native to Australia and commonly found in Uttaranchal Himalayas in between 1800-3000 m elevation. It is propagated by seeds or root cuttings.

Roots are having aromatic clove like odour and possess medicinally properties. The herb is astrigent, sudorific, cardial and febrifuge in nature.

Geum urbanum

Ginkago biloba Linn.

Family : Ginkgoaceae
Trade/Common Name : Maiden hair tree

A handsome, straight tree, upto 30 m high, spreadingly branched when young, bearing clusters of fan-shaped leaves. Flowers dioeciously.

It is native of China and occasionally cultivated in Indian gardens, particularly on hills. A few trees, deformed and multilated are found in H.P. It is propagated by seed and vegetative cuttings.

Leaves contain shikimic acid, flavonoids and also constitute phenolic components, sesquiterpene, bilobalide and beta-sterol. Seeds contain alkaloids – ginkgotoxin and amino-acid. Nuts are used against cancer in China. Leaves are used in cardiovascular disorders, increased cerebral blood circulation and parkinson's diseases. The seed is used as an anti-tusive and expectorant in Japan and China. It is an CNS improver, best brain tonic, memory sharpener in old ages.

Gloriosa superba Linn.

Family : Liliaceae
Trade/Common Name : Glory Lily

A beautiful, herbaceous, tall, glabrous, branching leaf tip climber, about 1-3 tall. Leaves alternative, opposite or terminally, whorled, lanceolate, strongly nerved, nerves parallel, leaf-tip ending in a tendril like spiral. Flowers large, showy, axillary, solitary on long pedicels, nodding and blooming in the month of September to continue October.

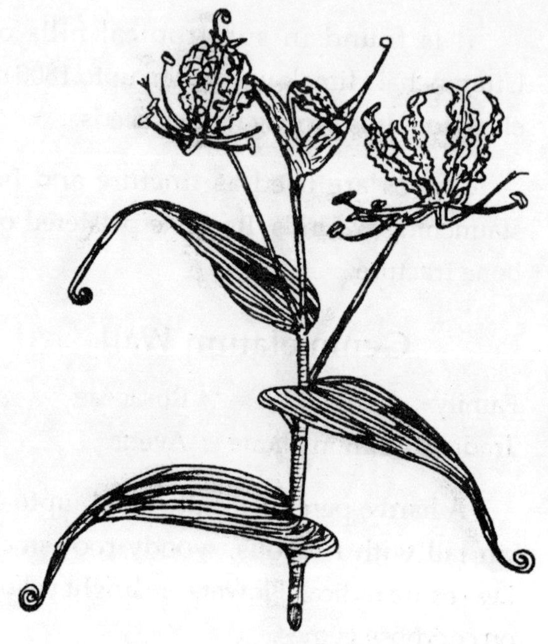

Gloriosa superba

It occurs in hedges and along forest boundries throughout tropical and sub-tropical India, upto 1500 m but frequently distributed in Sal forests. It is propagated by seeds.

It contains superbine, gloriosine and 1.3% colchicines. Tuber is pungent, bitter and acrid; heating anthelmintic, laxative, abortifacient, useful in chronic ulcers, leprosy, inflammation, piles, thirst etc. The leaf juice is used for killing lice in hair. Root paste is an antidote to snake bites, while root power is given in rheumatic fevers.

Glycyrrhiza glabra Linn.

Family : Fabaceae
Trade/Common Name : Liquorice

A perennial her with underground stem, highly branched attain upto 2 m height. The roots grey-brown exterior and yellow interior. Leaves alternate, pinnate, with 9-17 ovate, yellow green leaflets covered with soft hairs on underside. Flowers pea-like, lavender to purple in colour and appear in axil of terminal and axillary leaves in racemes.

It is distributed in Southern Europe, Syria, Iran, Afghanistan, Russia, China, Italy, Pakistan and Northern India. It is commercial cultivated as cash crop in semi-arid regions of Gujrat, Haryana and Rajasthan. It is propagated by seeds and stolons.

The liquorice roots contain glycyrrhizin, glycyrrhinic acid and glycyrrhetinic acid. It also consists of triterpenids and flavonoids and bitter principles. In Indian system of medicine, it is being used against sore throat, cough, myopia, stomach ulcers, stomach disorder, muscular pains, mouth disorders, baldness, irritation in urinary tract, corn etc. Its extract is used for flaviouring cigarettes, chewing and snuff tobacco.

Glycyrrhiza glabra

Gossypium hirsutum Linn.

Family : Malvaceae
Trade/Common Name : Short-Staple Cotton

Gossypium hirsutum

A shrub with palmately lobed leaves.

It is native to South America. It is cultivated and naturalized throughout tropics and subtropics. It is propagated by seeds.

Seeds yield an fatty oil. It is used in pharmaceutical industry as emollient, lubricant and as a vehicle for injections.

Grewia asiatica Linn.

Syn. *G. elastica* Royle.

Family : Tiliaceae
Trade/Common Name : Phalsa

A tree with densely crowed. Bark is grey coloured. Leaves are ovate, rouned and cordate. Flowers are yellow in colour and blooming in the month of April-May.

It occurs in Sal forests of India and commonly in Uttaranchal regions. It is propagated by seeds.

The bark and fruits have medicinal properties.

Grewia optiva Drum. ex. Burret

Syn. *G. oppositifolia* Roxb.

Family : Tiliaceae
Trade/Common Name : Bhimal

A deciduous, perennial tree, to 12 m high. Leaves ovate-lanceolate, acuminate, baserounded with prominent nerves,

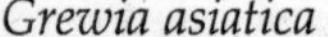

Grewia asiatica

Grewia optiva

margins serrate, stellately hairy on both surfaces. Flowers in axillary or leaf opposed cymes, 1-8 together; pedicel to 2 cm long, white in colour and blooming in the month of April-June.

It is commonly occurring at Arakot regions of Uttar Kashi between 500-1500 m elevation. It is propagated by seeds.

Bark is medicinally important. The extraction of bark is given to pregnant women for smooth delivery and also given in constipation.

Grewia sclerophylla Roxb. ex. G. Don.

Family : Tiliaceae
Trade/Common Name : Garbheli

A small shrub upto 2 m tall with scabrous bark. Leaves are ovate or acuminate. Flowers are pale-yellow and blooming in the month of July-August.

It is found in the sub Himalayan tract from Garhwal to Sikkim. It is propagated by seeds.

The fruit and bark are used as medicine.

Gymnema sylvestre (Retz) R. Br.

Family : Asclepiadaceae
Trade/Common Name : Merasingi

Gymnema sylvestre

Habenaria acuminata Linn.

Family : Orchidaceae
Trade/Common Name : Chandel

A terrestrial orchid upto 60 cm tall. Leaves are erect, linear-lanceolate, to 10 cm long and 1 cm broad. Flowers are green in colour, lip purple, trilobed and arranged elogate infloresence.

It is native to Annamalai hills, Kerala and Uttaranchal Himalaya. It is propagated by divisions of tubers.

The tubers are medicinally important and used as tonic.

Habenaria commelinifolia Wall. ex. Lindl

Family : Orchidaceae
Trade/Common Name : Gondla

A robust plant of 90 cm length. Leaves are oblong, finely acuminate to 15 cm long and 3 cm broad. Flowers are many, white in colour.

It is found in Chhatisgarh, Uttaranchal, Central India, Orissa and Karnataka. It is propagated by division of tubers

Roots are medicinally important. Locally, roots decoction is taken twice a day for one month in urinary trouble.

Habenaria goodyeroides Don.

Family : Orchidaceae
Trade/Common Name : Tengnoupal

A herb upto 60 cm tall with ellipsoid or oblong and hairy tubers. Leaves are elliptic upto 16 cm long and 7 cm broad. Flowers are sweet scented, olive green in colour.

It is found in Jharkhand, Bihar, Orissa, Sikkim, Manipur, Uttaranchal and Khasia hills at 1000 m altitude. It is propagated by tubers.

The tubers are used as tonic.

Habenaria griffithi Hook. f.

Family : Orchidaceae
Trade/Common Name : Dondonal

A terrestrial orchid with tuberous roots. Flowers are white, pinkish in colour.

It is found in Garhwal hills of Uttaranchal between 1500 to 2100 m elevation. It is propagated by divisions of tubers.

Tubers are used as tonic.

Habenaria intermedia Don.

Family : Orchidaceae
Trade/Common Name : Ridhi-Bidhi

A semi-tereta orchid upto 25 cm tall. Leaves are rounded at the base, long acuminate to 10 cm long and 5 to 7 nerved. Flowers are many, white or greenish in colour.

It is native to Kashmir to Kumaon (Uttaranchal), Sikkim and Khasia hills between 700-2300 m altitude. It is propagated by cuttings of tubers.

It is one of the eight drugs of the 'Ashtravargra', well known in Ayurveda. Tubers are used as general tonic. It is an ingredient of Chyavanprash.

Haematoxylon campechianum Linn.

Family : Fabaceae
Trade/Common Name : Logwood Tree

An ornamental thorny tree.

It is native to tropical America and now cultivated in India. It is propagated by seeds or cuttings.

Heart wood contains haematoxlin, used in dyeing. It is sudorific. It is used in medicine as a tonic and astringent, influxes and towards the end of dysentery.

Hedychium acuminatum Rose.

Syn. *H. spicatum* Bach. –Ham. ex Smith

Family : Zingiberaceae
Trade/Common Name : Van-Haldi

A perennial, rhizomatic, robust herb upto 1 m tall. The rhizomes are quite thick upto 7.5 cm in diameter, aromatic, knotty, spreading horizontally under the soil surface, grey brown in colour and have long, thick fibrous roots. Leaves large with long clasping shealths, oblong-lanceolate, acuminate. Flowers pinkish white in terminal spikes and blooming in the month of July to continue August.

Hedychium acuminatum

It occurs in the entire Himalayan region between 800-2800 m. In Uttaranchal, it is found in isolated patches elsewhere and frequent groups in Doon and Bhilagana catchment forest between 1600-2500 m. It is propagated by rhizome cuttings.

The rhizomes contain about 4% of an essential oil containing P-methoxy cinnamic acid, ethyl cinnamate, d-sabinine sesquiterpene and also contain starch (52%), organic acid, glycosides, ash (4.6%), fixed oil and odorous material. The rhizome is aromatic, anti-arthritic, appetizer, cardiac, stimulant, carminative, deodorant, dentrifuge, hair tonic, stomachic and tonic. It also useful in colic, cough and asthma; diarrhoea, dysentery, dropsy, headache, hair falling, liver complaints, rheumatism skin diseases and vomiting.

Helianthemum canadense (L.) Michx.

Family : Cistaceae
Trade/Common Name : Rock Rose

A hardy evergreen perennial hairy shrub growing upto 45 cm in height. Leaves are small upto 2.5 cm long. Infloresence is cincinnus. Flowers are yellow in colour. Flowers contain no honey and is homogamous, with sensitive stamens which move upwards when touched.

It is found in various parts of India. It is propagated by seeds or divisions.

Entire plant is useful when in bloom. The decoction of plant is used for curing diarrhoea, sore throat, skin diseases and syphilis.

Helianthus tuberosus Linn.

Family : Asteraceae
Trade/Common Name : Jerusalem Artichoke

A perennial herb upto 2 m in height. Rhizomes (tubers) swollen, much-branched, white or creamy. Leaves simple, fast green. Flowers are golden yellow in colour.

It is found in wastelands, waysides, orchards places of throughout sub-tropical and temperate regions and introdued as weed. At present, it is cultivated. It is propagated by tubers.

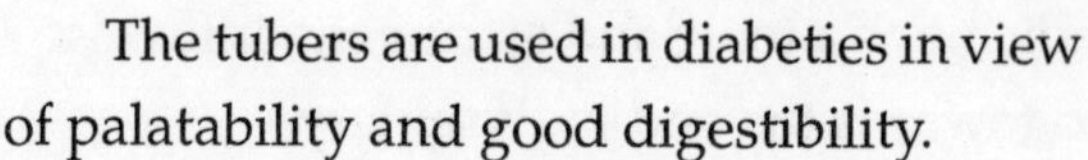

The tubers are used in diabeties in view of palatability and good digestibility.

Helicteres isora Linn.

Family : Sterculaceae
Trade/Common Name : East Indian screw tree

A sub-deciduous shrub or small tree with stem reaching a height of 1.5-4.5 m. Bark is grey in colour and in young parts is covered by stellate hairs. Leaves are simple, palmately nerved or obliquely cordate with serrated margin, scabrous above and pubescent beneath. Flowers are solitary or in sparse clusters with red petals turning pale blue when old and blooming in the month of may to continue August.

Helianthus tuberosus

Helicteres isora

The plant occurs often gregariously, throughout India, from Jamuna eastward to Nepal, Bihar, Jharkhand, Bengal and Southwards in Central, Western and Southern India and Andaman Islands. It is propagated by seeds.

Bark contains Saponins, sugar, phlobotannins, lignin, phytosterol, hydroxy-carboxylic acid and an organic-yellow crystalline colouring matters. The root and stem barks are considered to be expectorant, astringent and antigalactagogue. Bark is used in diarrhorea, dysentery, scabies, biliousness and is especially useful in gripping of the bowls. Juice extracted from roots is used in antidiarrhoeal and antidysenteric formulation. In tribal areas, the seeds powder (about 20 g) is used thrice a day for one week in the treatment of dysentery and stomach pain.

Hemidesmus indicus (L.) Schult.

Family : Asclepiadaceae
Trade/Common Name : Country Sarsaparilla

A prostrate or twining undershrub with aromatic rootstocks.

It is native to India and commonly occurs in West Bengal, Madhya Pradesh, Chhatisgarh, Rajasthan and Southern India. It is propagated by seeds.

Ethno-medicinally, the root decoction (about 10 ml) is used thrice a day for one month as blood purifier and skin diseases.

Hemidesmus indicus

Heracleum candicans DC.

Family : Apiaceae
Trade/Common Name : Hogweed

A perennial herb upto 2 m tall. Leaves 1-2 pinnate, radical ones pinnatified with large obtusely-lobes segments. Flowers white in large compound umbels, 12-45 ray and blooming in the month of June to continue upto September.

It is found in forest clearings, shrubberies, wayside and borders, cropfields along with streams in Western Himalayas from Kashmir to Uttaranchal. It is propagated by root cuttings.

Heracleum candicans

The roots are sweet smelling. It contains courmarins and used in the preparation of suntan lotions and possessing anti-leucodermal properties.

Heracleum lanatum Michx.

Family : Apiaceae
Trade/Common Name : Patrala

Rootstock stout, perennial, odorous, stem upto 1.75 cm tall. Densely pubescent. Leaves pinnate or pinnatifid, leaflet oblong, pinnately lobed and irregularly toothed, lower surface pubescent. Flowers small, white in compound umbels and blooming in the month of July to continue upto August.

It occurs globally from Pakistan to S.W. China 1800-4300 m in open slopes and drier area. The plant is frequently found to occur in hilly areas of Uttaranchal. It is propagated by seeds or divisions.

Roots are the chief source of xanthotoxin, a photosensitizing coumarin and the bergapten. The roots of the plant contain 6-9% of the total coumarins, the furanocoumarins, of which on hydrolysis and methylation, the yield of xanthotoxin is 1.5%. The roots is used in highly efficacious in the treatment of leucoderma and psoriasis. It is also used in the preparation of Sun-tan lotions.

Heracleum wallichii DC.

Family : Apiaceae
Trade/Common Name : Sikkim Herb

A perennial herb upto 1 m tall. Leaves pinnate, radical ones pinnatified with large obtusely-lobes segments. Flowers white in large compound umbels with numerous ray. The fruits come up in characteristic umbel during monsoon and is winged. These become dry by the onset of winter and ready for dissemination by wind.

It is found in moist along with streams in Sikkim Himalayas between 1500-2500 m elevation. It is propagated by seeds and root cuttings.

The roots are sweet smelling. It contains courmarins and used in the preparation of sun tan lotions and possessing anti-leucodermal properties. Locally, the dried fruits are taken orally during influenza.

Herniaria glabra Linn.

Family : Caryophyllaceae

Trade/Common Name : Burst Wort

A hardy trailing perennial herb. Foliage is mass-like, forming mats which turn deep bronze red in winter. Leaves are dark green in colour and 2.5 cm long. Flowers are apetalous.

It is native to Europe and introduces in India as ornamental plant for its attractive foliage. It is propagated by seeds or by divisions.

It is used as a stomachic and also recommended for urinary diseases.

Herniaria glabra

Hibiscus rosa-sinensis Linn.

Family : Malvaceae

Trade/Common name : China Rose

A shrub, up to 3 m tall. Leaves ovate, glabrous. Flowers solitary, axillary, crimson, never set fruit.

It is commonly planted as an ornamental in house gardens and also found throughout the India. It is propagated by shoot cuttings.

Hibiscus rosa-sinensis

The flowers and leaves are used against a variety of ailments such as alepacia, burning sensation in the body, diabetes, menstrual disorder and piles. The flowers are considered demulcent, emollient, refrigerant, aphrodisiac and emmenagogue. Decoction of flowers given in bronchial catarrh. Leaves emollient and aperients. Fresh root juice is given for gonorrhoea and powdered root for menorrhagia.

Hippophae rhamnoides Linn.

Family : Elaeagnaceae
Trade/Common Name : Seabuck thorn

A varying sized shrub, shoot often modified into spines. Leaves lanceolate-linear, obtuse, with peltate and stellate scales on lower surface. Male flowers in clusters at the bases of the shoot; female flowers pediellate in clusters red or orange colour and blooming in the month of June to continue upto August.

The plant occurs from Pakistan to Himachal Pradesh and Central Asia between 2100-3600 m on river sides. It is propagated by seeds.

The plant contains harmol, Harman, polyphenols, quercetin and isorhamnetin. Sertonin has been extracted from its bark. Fruits are rich in vitamin–C. Fruits are used against pulmonary complaints. It is used as emollient in prevention of eye region skin wrinkles and in other cosmetic preparation.

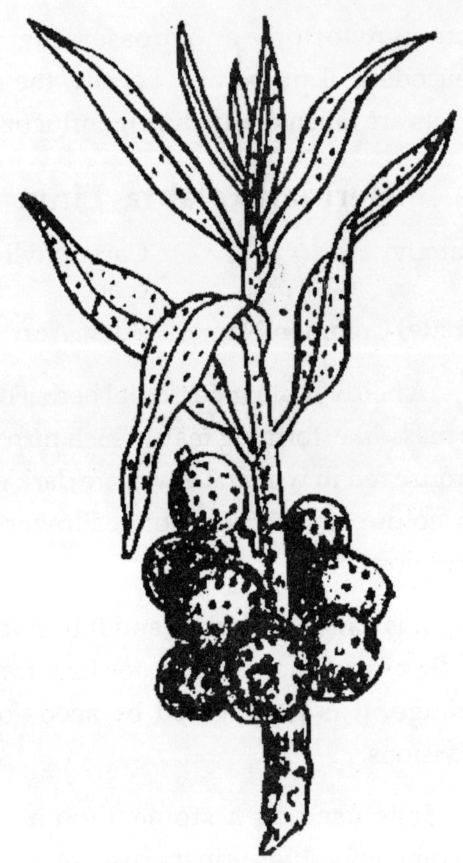

Hippophae rhamnoides

Hippophae salicifolia D. Don.

Family : Eleagnaceae
Trade/Common Name : Seabuck thorn

An erect shrub or a small tree. Leaves long, linear lanceolate, softly white-pubescent above when young, densely tomentose or stellately-hairy beneath; margins recurved; flowers are blooming in the month of June-August.

The plant occurs from Pakistan to Himachal Pradesh and Central Asia

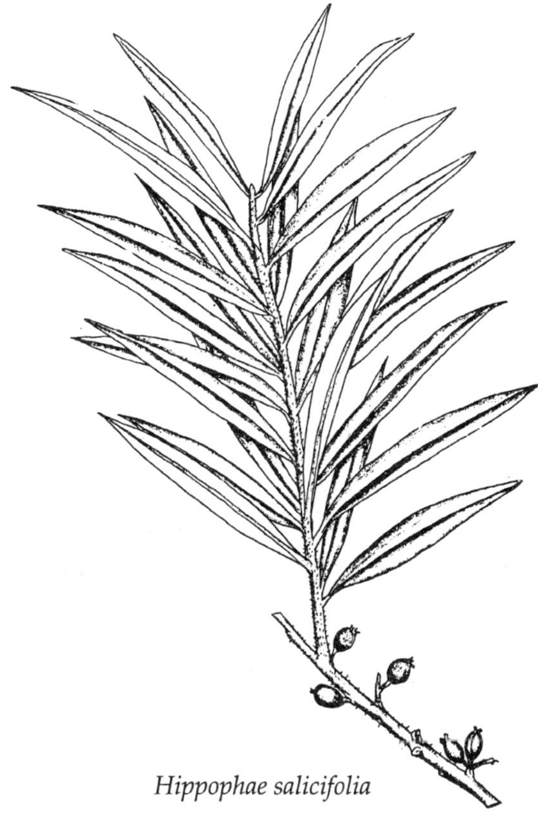

Hippophae salicifolia

between 2100-3600 m on river sides. It is propagated by seeds.

Bark contains a-sitosterol and bark has given promising results as a cure for certain types of cancer. The paste of its bark is used for curing wounds and ulcers. Plant is a medicine in tumors, used in sun-burn preventing, medicinal and cosmetic preparations for skin.

Holarrhena antidysenterica Wall. ex DC.

Syn. *H. pubescens* Wall ex G. Don.

Family : Apocynaceae
Trade/Common Name : Kurchi

A tall shrub or a small deciduous tree with exfoliating bark. Leaves are stalked, opposite, ovate-oblong, entire and pale-green in colour. Flower are white in colour and are fragrant and blooming in the month of May to contiune upto June.

The plant occurs almost throughout India upto an altitude of 1250 m often gregariously in deciduous forest and open waste lands and especially abundant in the sub-Himalayan tract. It is propagated by seeds.

Bark contains conessine, kurchicine, kurchine, conamine, conkurchine, halorrhenine etc. bark is used in dysentery and is prescribed in piles, anaemia, asthma, bronchopneumonia, dropsy, dysurea, influenza, rheumatism, tooth-ache, vomiting and nausea, intestinal worms dyspepsia, chest complaints and n the diseases of the skin and the spleen. Seeds

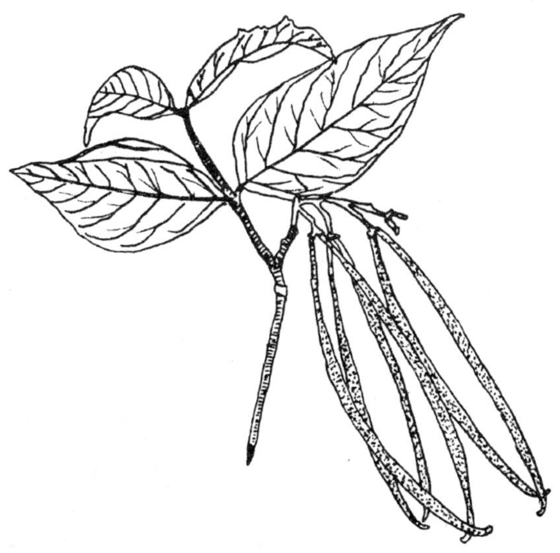

Holarrhena antidysenterica

are astringent, carminative, febrifuge, antidysenteric and antithelmintic. It is antidote to poison, anti-periodic, aphrodisiac, astringent bitter, carminative, conceptive, cooling, digestive, expectorant, febrifuge and tonic. Locally, its bark powder is orally administered with water during amoebic dysentery.

Humulus lupulus Linn.

Family : Moraceae
Trade/Common Name : Hops

A perennial, twining, dioecious scabrid herb. Aerial stems are oppositely branched, angular. Leaves are opposite, stipulate; lower larger, 3-5 lobed; upper smaller; ovate or cordate, serrate. Male flowers are small, greenish yellow, arranged in paniculate cymes. Female flowers are in a cone like inflorescence resulting into scaly bracteates fruiting bodies and blooming in the month of June-July.

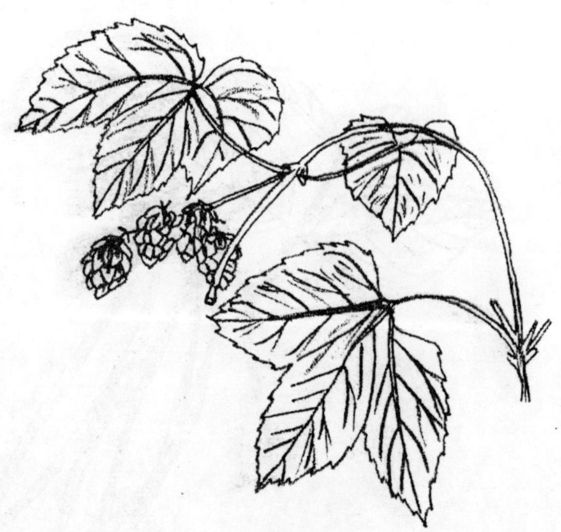

Humulus lupulus

The plant is native of North America and cultivated in commercial scale in hills of Himachal Pradesh. It is propagated by seeds or vegetative methods.

Mature female cone contains essential oil, choline, asparaginer and lupulin and also a bitter substances like lupulones, humuliones and hulupones. It is used as tonic, stomatic and diuretic. It prevents worms and allays the nervous tension and prevents indigestion. Infusion is regarded as tonic and sedative.

Hydnocarpus kurzu (King) Warb.

Family : Flacourtiaceae
Trade/Common Name : Spongy Berry Tree

A tree of 15 m tall or more with drooping branches. Leaves are leathery, pointed at the tip and upto 20 cm long. Flowers appear in small axillary clusters, small, yellow in colour. Fruit are round, 7 cm across, brown in colour, with many seeds.

It is native to evergreen forests of Assam and Tripura. It is propagated by seeds or cuttings.

The seeds yield an oil, used in certain preparations, are injected for the treatment of leprosy. The bark contains tannins and is used to cure fever.

Eupatorium cannabinum

Euphrasia officinalis

Euphorbia hirta

Foeniculum vulgare

Fritillaria roylei

Gloriosa superba

Glycyrrhiza glabra

Hemidesmus indicus

Heracleum candicans

Holarrhena antidysenterica

Hyoscyamus niger

Inula racemosa

Jasminum multiflorum

Jatropha curcas

Juglans regia

Juniperus macropoda

Jurinea macrocephala

Justicia adhatoda

Lagotis cashmeriana

Lannea coromandelina

Lavandula officinalis

Lavatera cashmeriana

Lawsonia inermis

Linum usitatissimum

Mallotus philippinensis

Mangifera indica

Marrubium vulgare

Matricaria chamomilla

Hydnocarpus laurifolia Sleumber.

Syn. *H. weightiana* Blume.

Family : Flacourtiaceae

Trade/Common Name : Yellowberry Tree

A tree of 15 m tall or more with branches. Leaves are leathery, pointed at tip. Flowers appear in small axillary clustered, small, yellow in colour.

It is native to China, Southern Asia and Western India. It is found in evergreen forests of Western Ghats of India. It is propagated by seeds.

Seeds yield an volatile oil. The oil contains cyclopentenyl monocarboxylic acid, chaulmoogric acid and garlic acid. The oil is used to treat leprosy, chronic skin diseases and rheumatic pains, strains, tubercular laryngitis and inflammations.

Hydnocarpus laurifolia

Hygrophila auriculata (Schum.) Heine.

Family : Acanthaceae

Trade/Common Name : Talmakhana

An underbranched stout erect herb upto 1.5 m tall. Stem are four angled and hairy. Leaves are whorled, 6 at a node, each possses sharp yellow coloured spines on its axil. Flowers are purple blue about 3 cm long, 2 lipped, lower lip folded, 8 in a whorl.

It is found in marshy, moist regions and throughout India. It is propagated by seeds.

Seeds and roots are diuretic. Leaves are used to treat cough and urethral discharges. Seeds are also employed in the treatment of veneral diseases. The drug from plants is used to cure diseases of urinogenital system, jaundice, dropsy and rheumatism.

Hygrophila auriculata

Hyoscyamus muticus Linn.

Family : Solanaceae
Trade/Common Name : Egyptain Henbane

A perennial herb upto 1 m tall. Flowers are funnel shaped, yellow or greenish white. Seeds are kidney shaped.

It is native to Egypt, Sudan, Syria and Afghanistan. It is cultivatedin various parts of India. It is propagated by seeds and cuttings.

The total alkaloids content ranges from 0.5 to 1.5%. The leaves and flowering tips contain hyoscyamine and trophane alkaloids. It is used as poison, as narcotic, sedative and antiseptic and mydriatic. The leaves are alos smoked for intoxication.

Hyoscyamus muticus

Hyoscyamus niger Linn.

Family : Solanaceae
Trade/Common Name : Khurasani ajwain

An erect, viscidly hairly, foetid annual or biennial upto 1.5 m tall. Leaves are redial and cauline, coarsely dentate, to pinnately lobed. Flowers are yellowish green, having dark purple centers, sessile or sub-sessile and are in terminal scorpioidal cymes and blooming in the month of July-August.

The plant occurs in wastelands, rubbish heaps, forestclearings in the Western

Hyoscyamus niger

Himalayas from Kashmir to Uttaranchal at altitude of 1600-3000 m. It is propagated by seeds.

The herb contains total alkaloiks: Roots (0.16%), leaves (0.045-0.08%), flowering tops (0.07-0.1%) and seeds (0.06-0.1%). Leaves and flowering tops contain principal alkaloids hyoscyamine, hyoscine or scopolamine, volatile bases; a bitter glucoside, hyoscypicrin, choline, mucilage etc. It has anodyne, narcotic and mydriatic properties. It is principally employed as a sedative in nervous affections and irritable conditions such as asthma and whooping cough.

Hypericum perforatum Linn.

Family : Hypericaceae
Trade/Common Name : St. John's Wort

A perennial herb. Stem 30-60 cm tall, 2-angled. Leaves oblong; obtuse veins pellucid, lower surface pale. Flowers in terminal corymbs and blooming in the month of April to continue upto August.

It occurs sporadic in shady, waysides in the forest clearings, grazing ground, damp forest of Western Himalaya at altitude of 1000-2600 m. It is propagated by seeds.

Its aerial part contain anthraquinone derivatives, hypericin, usigtoercin and protohypericin, phenolic acids and tannins. An oil (oleum hyperici), obtained by infusing fresh flowers in olive oil, is used externally for sores, wounds, ulcers, swelling and sometimes rheumatism and lumbago. Herb possess sedative and astringent properties. It has been used for excitability neuralgia, fibrrositic, sciatica, wounds and specifically for menopausal neucrosis.

Hypericum perforatum

Hyssopus officinalis Linn.

Family : Lamiaceae
Trade/Common Name : Hyssop

An aromatic, shrubby perennial, 30-60 cm high. Branches are erect or diffuse and the leaves are sessile, linear, oblong or lanceolate, obtuse and entire. Flowers are blue purplish in colour, aggregated in

Hyssopus officinalis

axillary tufts and are arraged unilaterally on terminal branches and blooming in the month of July-September.

It occurs in the Himalayan region from Kashmir to Uttaranchal at altitude of 2600-3500 m. It is propagated by seeds or division of stems.

Leaves and flowering tops yield an essential oil (1%). The volatile oil consists of the betone 1-pinocamphene, â-pinene, L-pinene, camphene, 1-pinocamphenol etc. Besides the volatile oil, the herb contains fat, sugar, choline, tannin, carotin and xanthophylls. The herb is considered as stimulant, carminative and pectoral and used in such medicine. The crushed herb is applied as a resolvent ad vulnerary. It is recommended for use as a gargle in throat inflammations and hoarseness and internally for chronic bronchitis and bronchial asthma.

Inula recemosa Hook. f.

Family : Asteraceae
Trade/Common Name : Pushkarmool

A stout herb, upto 1.5 m tall. Leaves are leathery, rough above, densely hairy beneath ad toothed. Radial leaves are long,

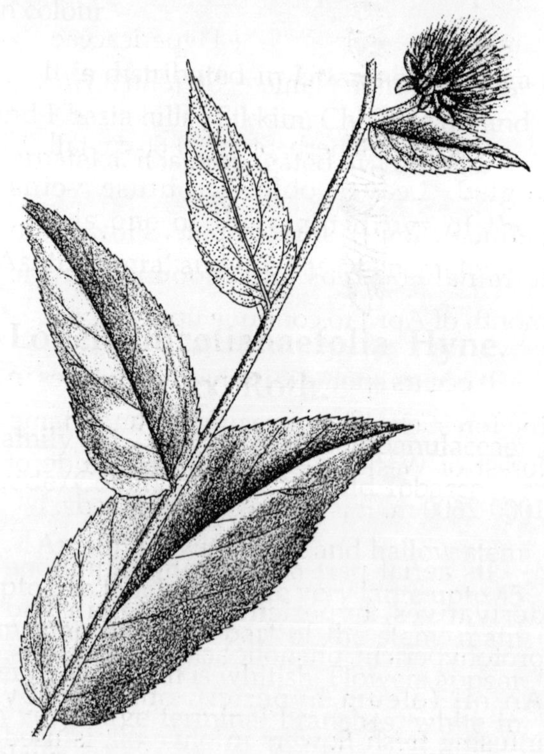

Inula recemosa

stalked, elliptic, lanceolate; cauline leaves are ovate-oblong semi-amplexicaule, often deeply lobed at the base. Flower heads are many, yollow in colour and arising in terminal racemes and blooming in the month of July to continue upto August.

It is found in cultivated areas, forest clearings, shrubberies in the North-Western Himalayas from Kashmir to Uttaranchal at altitude ranging from 1600-4200 m. It is propagated by divisions of roots or shoots.

The root yields an inulin and an essential oil containing iso-alantolactone and di-hydralactone. Roots possess strong anthelmintic properties. It produes fall in blood pressure and stimulates peristaltic movements of intestine. Alantolactone is antiseptic, expectorant and diuretic. The seeds are bitter and aphrodisiac. Root extract acts as anti-inflammatory, antipyretic/ antiseptic and antiasthmatic.

Ipomoea hederacea Auct. Non Jacq.

Family : Convolvulaceae
Trade/Common Name : Pharbitis seeds

A spreding herbaceous twining plant. Stem are retorsely covered with hairs. Leaves ovate-cordate, 3 lobed. Flowers in axillary stout peduncled, upto 5 cm long, blue coloured tinged with pink. After plucking, flowers become purplish in colour.

It is widely distributed in tropical and warm temperate regions of the world. It is found throughout India upto 1800 m elevation. It is propagated by seeds or stem cuttings.

Ipomoea hederacea

Seeds are medicinally important. It contains resinous matter, affixed oil, small amount of saponin, mucilage and tannin. Dried seeds are used as purgative. In over doses the drug cause symptoms of irritant poisoning. It is also used in medicine as extract, a tincture and a compound powder.

Iris kumaonensis Wall. ex. Don.

Family : Iridaceae
Trade/Common Name : Lhathum

A perennial herb with creeping rhizome. Flowers are blue and blooming in the month of May to continue upto July.

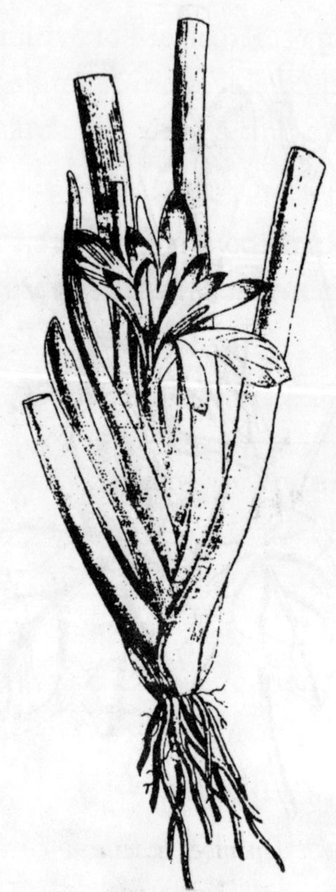

Iris kumaonensis

It occurs frequently in th temperate zones on open grassy slopes in Northern Himalayas between 3000-4000 m elevation. It is propagated by corms and seeds.

Roots are used against kidney trouble and fever.

Ixora alba Hort.

Syn. *I. Chinensis* cv. alba

Family : Rubiaceae
Trade/Common Name : Leopard Lily

A compact, bushy shrub. Leaves are dark green in colour; verticular, coriaceous, petiolate and sessile. Flowers borne in corymbs and white and sweetly scented.

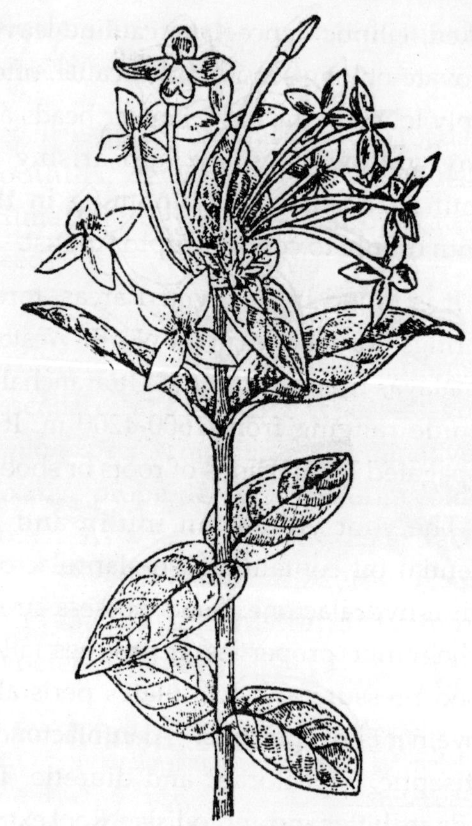

Ixora alba

It is a native of India. It is found in Southern India. It is propagated by seeds, layering or cuttings.

Flowers and leaves are used for medicinal purposes.

Jasminum angustifolium Vahl.

Family : Oleaceae
Trade/Common Name : Ban Mallika

A dwarf climbing shrub. Leaves are small, ovate, smooth, rounded at the base. Cymes are 1-3 flowered, produces white star like fragrant flowers during winter and spring.

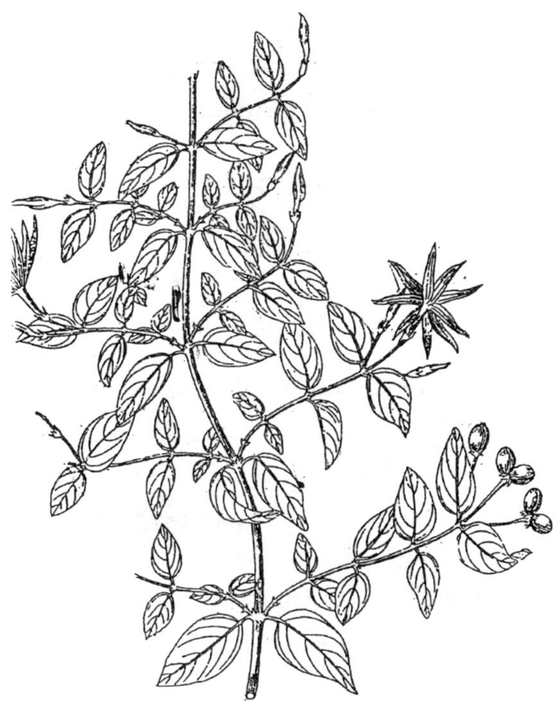

Jasminum angustifolium

It is native to India. It is found in Deccan Paninsula, South Travancore. It is propagated by suckers.

Flowers are used for making garlands, adorning of hairs by women, religious and ceremonial functions, for perfumery and for the production of essential oil.

Jasminum arborescens Roxb.

Syn. *J. latifolium* Roxb.; *J. roxburghianum* Wall.

Family : Oleaceae

Trade/Common Name : Tree Jasmine

A large shrub or shrubby tree with hairy branchlets. Leaves are ovate, acute with petiole. Cymes branching into three divisions, hairy, not dense, usually 12-20 flowered. Flowers are fragrant white, blooms in winter through summer.

Jasminum arborescens

It is native to India. It is inhabiting in tropical north-west Himalayas from Terai of Oudh and Kumaon, Deccan Peninsula, altitude 200-1000 m, Tirhut division of Bihar, Jharkhand, Orissa. It is propagated by suckers.

Flowers are used for making garlands, adorning of hairs by women, religious and ceremonial functions, for perfumery and for the production of essential oil.

Jasminum auriculatum Vahl.

Syn. *J. affine* Wt..; *J. ovalifolium* Wt.

Family : Oleaceae

Trade/Common Name : Joohi

A climbing shrub, branchlets and leaves are grey-pubescent. Leaves are trifoliate, opposite with a pair of minute lateral leaflets or sometimes absent. Cymes are compound, many flowered, terminal, paniculate, grey pubescent, produce numerous star shaped white flowers from spring to summer and also in rains.

It is native to India. It is distributed in Deccan Peninsula, North West India, West Bengal. S. Travancore and Western Ghats. It is propagated by suckers.

Flowers yields an essential oil (0.29-0.34%). It is used for making garlands, adorning of hairs by women, religious and ceremonial functions, for perfumery and for the production of essential oil.

Jasminum auriculatum

Jasminum calophyllm Wall.

Syn. *J. courtallense* Wt.

Family : Oleaceae

Trade/Common Name : Herbini

A climbing shrub. Leaves are opposite, smooth, leaflets elliptic or ovate, acute or caudate-acuminate at apex, rounded or obtuse or shortened at base, up to 12 cm by 5 cm. Cymes are compound branching into three divisions, 10-15 flowered. White fragrant flowers appear during February to June.

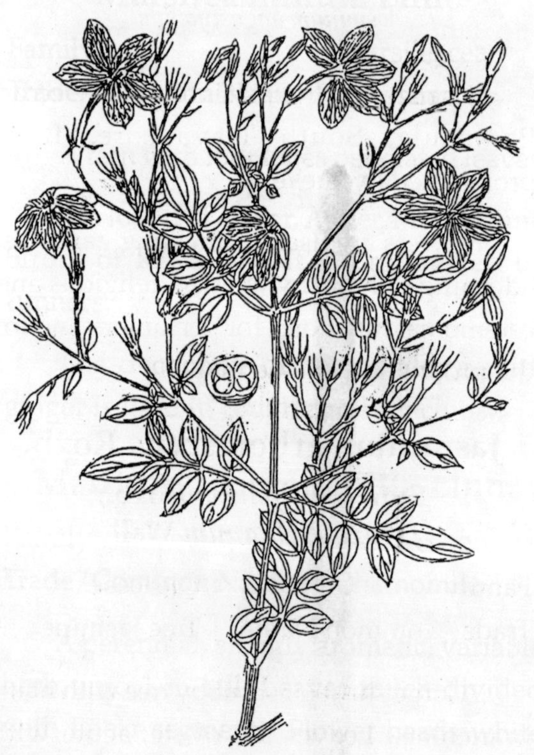

Jasminum calophyllm

It is native to India. It is found in South Deccan Peninsula, Nilgiri Hills, Western Ghats, Annamalai and Tinnevelly Hills; altitude up to 1300 m. It is propagated by suckers.

Flowers are used for making garlands, adorning of hairs by women, religious and ceremonial functions, for perfumery and for the production of essential oil.

Jasminum humile Linn.

Family : Oleaceae
Trade/Common Name : Yellow jasmine

An erect, diffuse shrub with profuse, angular branches. Leaves are alternate, 5-10 cm long, leaflets rae 3-7 in number and are oblong or ovate. Flowers are yellowish in colour in terminal corymbose cymes and blooming in the month of September-November.

It is common in the outer Himalayas (Uttaranchal) at 1200-2800 m altitude. It is also found throughout the sub-tropical Himalayas from Kashmir to Nepal upto an altitude of 2800 m and in Mount Abu and the hills of South India above 1500 m. It is propagated by seeds.

Flowers yield a volatile oil, whereas the leaves yield salicyclic acid. The paste of leaves is applied locally in skin diseases, wound and ulcers. Leaves are chewed in stomatitis and toothache. It is also considered as an antidote in the poisonous cases, besides being used as an astringent, tonic to the heart and bowels. Useful in the treatment of skin eruptions, wounds and ulcerations.

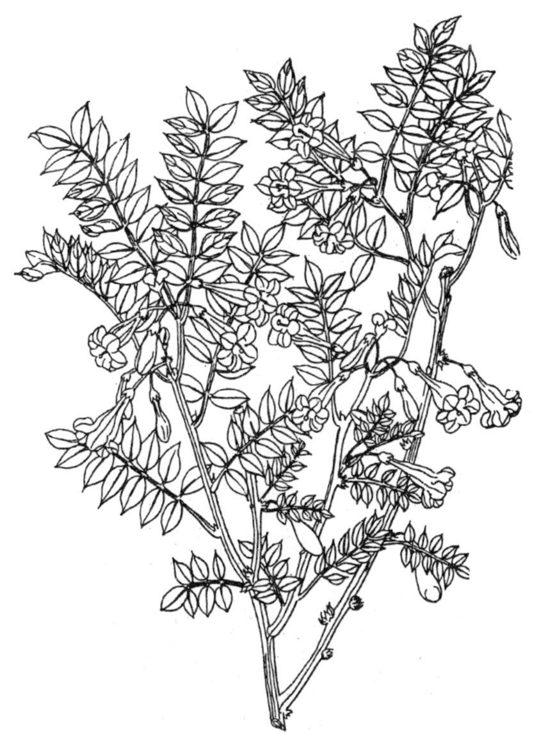

Jasminum humile

Jasminum mutliflorum Roth.

Syn. *Nyctanthes mutliflora* Burm. f.

Family : Oleaceae
Trade/Common Name : Safed Chameli

A bushy, densely hairy shrub. Leaves are simple, opposite, ovate, acuminate, more or less pubescent beneath. Cymes are terminal, many flowered, umbel. Flowers are complete white, slightly fragrant, sun-sessile, blooms from December to August.

It is native to India. It is found throughout India, common in Sub-

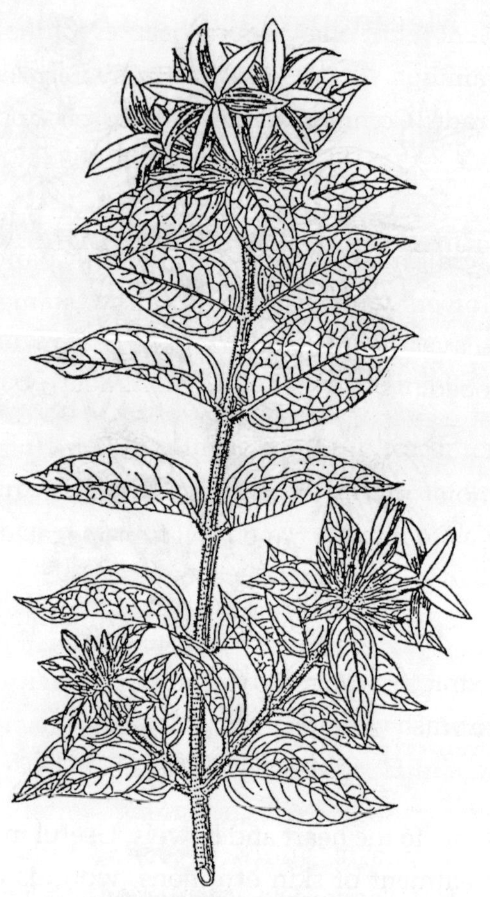

Jasminum mutliflorum

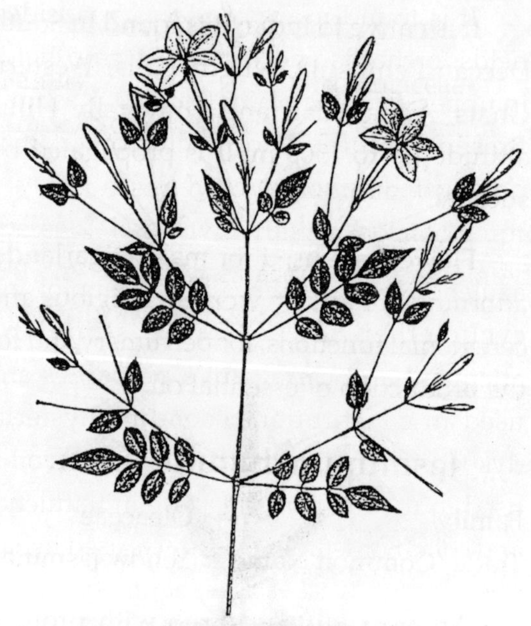

Jasminum officinale

Himalayan tract and moist forest of Western Ghats; altitude up to 1700 m. It is propagated by suckers.

Flowers are used for making garlands, adorning of hairs by women, religious and ceremonial functions, for perfumery and for the production of essential oil.

Jasminum officinale Linn.

Family : Oleaceae
Trade/Common Name : Sweet Jasmine

A large climbing shrub with long weak dark green branches. Leaves are terminal 1-10 flowered, often leafly; bracts narrow to 0.8 cm long. Flowers fragrant white, appear in winter seasons.

It is native to India. It grows wild from Kashmir to Iran. In India, it is occurring in Kashmir, sub-tropical North-West Himalayas; altitude 700 to 1700 m. It is propagated by suckers.

Flowers are used for making garlands, adorning of hairs by women, religious and ceremonial functions, for perfumery and for the production of essential oil.

Jatropha curcus Linn.

Family : Euphorbiaceae
Trade/Common Name : Barbedos Nut

A deciduous shrub or small tree upto 5 m height. Yound shoots glandular tomentose, with saponaceous juice; bark grey, smooth. Leaves alternate, orbicular-cordate or 3-5 palmately lobed, base cordate, apex acute or obtuse. Flowers yellowish-green, in terminal and sub-terminal, long, cymose, panicle.

It is native of tropical America. It is found widely in waste-lands, margin land and dry exposed slopes of lowers hills upto 1000 m elevation. It is propagated by seeds.

Seeds contain toxic principle curcin. Seeds oil medicinal as anthelminitic and applied externally in rheumatism ae well as skin disorders. Seeds in over doses regarded as poisonous. Locally, juice of the plant is applied in eczema, ringworm and useful in scabies.

Juglens regia Linn.

Family : Juglancaceae
Trade/Common Name : Walnut

A handsome, tall tree with imparipinnate leaves. Leaflets are in pairs, sessile, oblong, rounded at base, acuminate. Bark is flat or in curved pieces, dark gray in colour and smooth from inside. Fruit is a fleshy drupe, oblong, ovoid and of olive green colour, when immature.

The tree is found growing wild throughout the Himalayas and hills of Assam at an altitudes of 1000-3300 m and is generally cultivated in hills. It is propagated by seeds.

Jatropha curcus

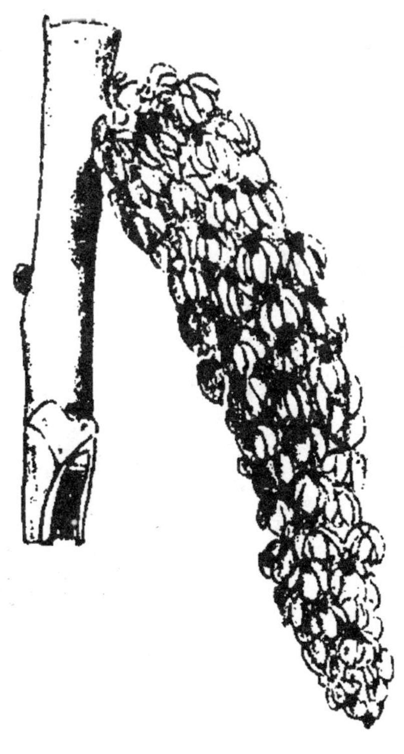

Juglens regia

Root bark contains juglandic acid, juglonone. Fruits contain the oxalic acid, resin. Karnils contain B group of vitamins and several mineral elements. The leaves are alternative and given in scrofula, rickets and leucorrhoea. The decoction of the bark is galactafuge and is used to stop mammary secretions. It is used as an astringent to check diarrhoea and menorrhagia and for gargling in case of a sore throat.

Juniperus communis Linn.

Family : Cupressaceae
Trade/Common Name : Juniper

A dense, dwarf, procumbent, bushy, evergreen shrub (60-150 cm) with upright branches, or a small tree reaching upto 9 m height. Bark is chocolate brown, peeling off in papery shreds; leaves numerous and in whorls of 3, persistent, spreading at right angles and rigid; young twigs are divaricated; Flowering in the month of March-April.

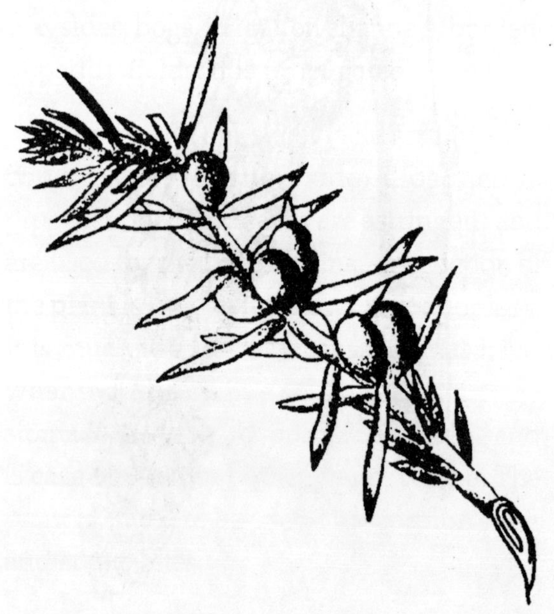

Juniperus communis

It occurs in dry slopes, gravely substrate, exposed plateaus in the temperate regions from Kashmir to Uttaranchal at the altitude of 3000-4500 m. It is propagated by seeds, cutting and layers.

It yields an essential oil (0.25%). It also contains resin (10%), a bitter substance, juniperine and organic acids. Juniper is aphrodisiac, hemicrania, chronic, bronchitis, diseases of liver and spleen. Its oil possess carminative, stimulant and diuretic properties. Wood is sudorific in its action and is a substitute for guaiacum and sassafras.

Juniperus macropoda Boiss.

Family : Cupressaceae
Trade/Common Name : Indian Juniper

A tree upto 18 m tall with light bark, open foliate with spreading, sharp pointed leaves on lower braches. Scale leaves are on the upper branches and branchlets. Flowers are blooming in the month of May-August.

The plant occurs in dry rivers valleys, gregarious forming open forests from Pakistan to Uttaranchal at the 2400-4300 m altitude. It is propagated by seeds.

Berries yield essential oil constituting sugiol, junipodin and its glycosides junipin. It is useful in colic, cough and pectoral

Juniperus macropoda

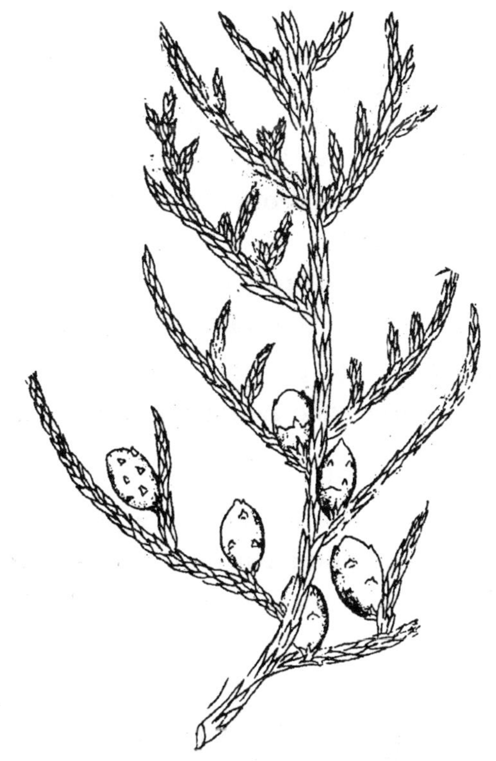

Juniperus recurva

affections, diarrhoea, impotency, indigestion, leucorrhoea, paralysis, piles and skin affections. Fruit and resin used in ulcers and nasal polyps.

Juniperus recurva Bach. Ham.

Syn. *J. squamata* Buch.-Ham.

Family : Cupressaceae
Trade/Common Name : Doop

A prostate or decumbent shrub or a small tree of graceful, drooping habit, often attaining a height of 10-12 m. Stems creep over ground and rock boulder. Leaves owl shaped; flowers monoecious or dioecious and blooming in the month of June-August.

It is distributed throughout the temperate and alpine zone from Kashmir to Uttaranchal and Assam at altitudes of 2200-4500 m. It is propagated by seeds, cutting and layers.

Wood yields an resin and fruits give an essential oil (0.46-0.88%). It is suitable for use as a pencil wood. The smoke from the green wood is reported to be emetic. Ethyl alcohol extract of aerial parts has been reported to be anticancer.

Jurinea dolomiaea Boiss

Syn. *J. macrocephala* (Royle) Clarke.

Family : Asteraceae
Trade/Common Name : Dhoop

A prostrate, perennial herb with a central dome shaped cluster of large purple flower heads upto 10 cm across. Root is stout, softly, aromatic, dark brown, perennial, going deep into the soil. Leaves are radical, spreading in a rosette like manner, hairy above, thickly white tomentose beneath; pinnally divided into broad, lobulate, toothed segments. Heads 3-30, purple, short stalked, upto 4 cm aggregated in the center of the leaf rosette and blooming in the month of July continue upto September.

It is commonly found in grasslands, open slopes, alpine pastures/slopes from Kashmir to Uttaranchal between 3000-4300 m at altitudes. It is propagated by seeds or divisions of roots.

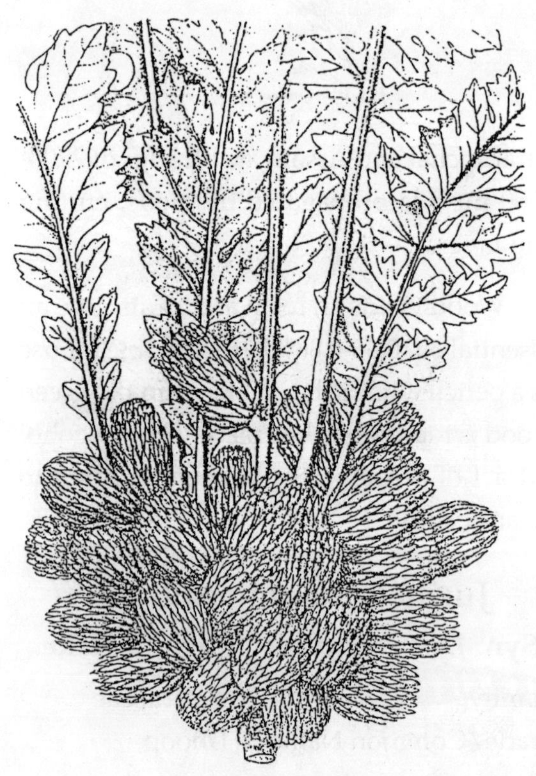

Jurinea dolomiaea

The roots contain an aromatic resin and used as incense and form a chief ingrediant of doop industry. The roots are considered stimulant and given in fever after child birth. Roots decoction given in colic. Bruised roots applied to eruptions.

Justicia adhatoda Linn.

Syn. *Adhatoda vasica* Nees.

Family : Acanthaceae
Trade/Common Name : Vasa, Vasaka

An evergreen shrub with yellowish brown bark. Leaves are elliptic, acuminate; base narrowed, minutely pubescent, especially when young; lateral nerves, 8-14 pairs. Flowers arise in dense, axillary pedunculate spikes and blooming in the month of December continue upto April.

It is found throughout in plains of India, and in the Sub Himalayan tracts upto 1450 m. It is propagated by seeds.

Leaves and flowers contain alkaloids vasicine. vasicinone, a-sitosterol, kaemp-ferol have been isolated from plant and vasicoline has beeen oxidized to vasicinone. Leaves and roots are useful in cough, asthma, bronchitis, rheumatism and as insecticidal. Flowers, leaves and roots are antiseptic, antispasmodic, allevative, blood purifier and tonic. It is common remedy for cough, lung complaints, anaemia and asthma. For immediate expectoration in children and spasmodic attack of asthma, fresh juice gives immediate relief, which may be followed by other preparations of the drug. The twigs of the plant when used as tooth sticks offer a cure for pyorrhea.

Kaempferia galanga Linn.

Family : Zingiberaceae
Trade/Common Name : Kapur kachuri

A herbaceous plant.

It is a tropical plant. It is found throughout India. It is propagated by seeds.

Rhizome yield an essential oil. It is used in indigenous medicine as stimulant, expectorant, diuretic and carminative. A decoction of rhizome is prescribed for dyspepsia, headache and malaria. Roasted rhizomes are applied hot in rheumatism and for hastening the ripening of inflammatory tumours. The leaves are used in lotions and poultices for sore eyes, sore throat, swelling, rheumatism and fever. The oil is also used in perfumery industries and curry flavourings.

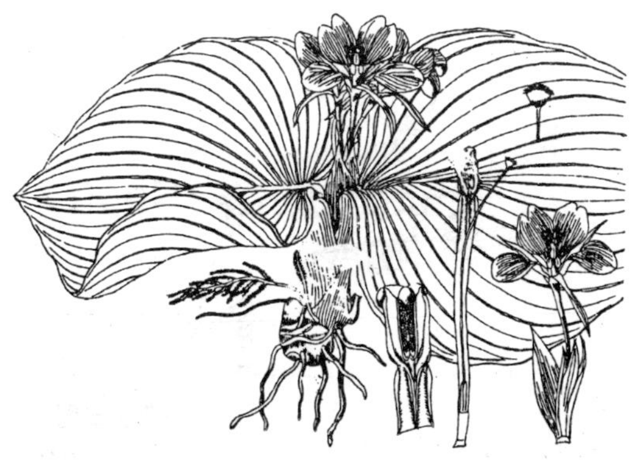

Kaempferia galanga

Kydia calycina Roxb.

Family : Malvaceae
Trade/Common Name : Puli

A deciduous, quick growing tree, 15-20 m tall. Leaves suborbicular or ovate-rounded, subcordate at base, acute or obtuse at apex, entire or irregularly serrate. Flowers in axillary or terminal panicles, white or pink in colour and blooming in the month of July-September.

It is distributed in tropical Himalayas and south-wards to Peninsular India. It is propagated by seeds.

Leaves are applied as paste in body pain. It also chewed for the formation of

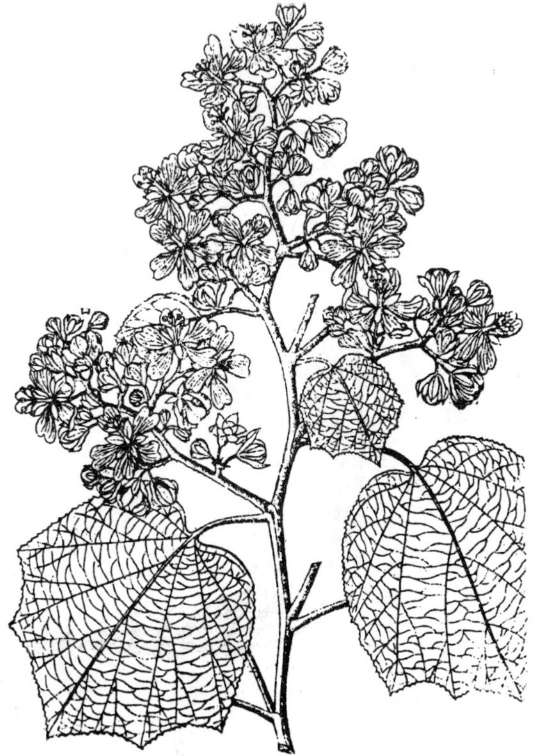

Kydia calycina

saliva. Locally, its poultice is used for skin diseases.

Lagerstroemia speciosa (L.) Pers.

Family : Lythraceae
Trade/Common Name : Queen Crape Myrtle

A showy flowering tree. Leaves are usually set in opposite pairs in two rows and have entire margins. Flowers are wrinked or crisped.

It is found in Southern and Eastern Asia. It is propagated by seeds or layering.

Roots are used as astringent, stimulant and febrifuge. The bark and leaves are purgative and seeds are narcotic.

Lagerstroemia speciosa

Laggera alata (D. Don) Schulz.

Family : Asteraceae
Trade/Common Name : Sapbuti

A herbaceous plant with purplish or pink flowers.

It is native to India and occurring at Chatishgarh and also found in tropical Himalayas. It is propagated by seeds.

Ethno-medicinally, leaf and root powder (about 30 g) is taken thrice a day for one week in snake bite.

Lagotis cashmeriana (Royle) Rupr.

Family : Scrophulariaceae
Trade/Common Name : Kashikhana

A small herb with oblong-elliptic leaves and dense cyclindrical spikes of dark blue tubular flowers.

It is found in moist slopes in alpine regions of western Himalaya from Kashmir to Himachal Pradesh. It is propagated by seeds.

Roots and leaves are bitter and used in fever and dyspepsia.

Lannea coromandelica (Houtt.) Merr.

Family : Anacardaceae
Trade/Common Name : Wodier

A deciduous tree with thick, soft branches; bark pale-brown, rough, exfoliating in thin round plates. Leaves alternate, imparipinnate, 25-40 cm long; leaflets 7-11, shortly petiolulate, ovate or ovate-oblong. Flowers unisexual, 4-merous, greenish-yellow, appearing before the leaves, male flowers in dense cymose-fascicles or compound drooping racemes; female flowers in simple racemes, crowded towards the tip of branches and blooming in the month of March-April.

It is found in the lower valleys of Uttaranchal Himalayas and also at Chatishgarh. It is propagated by seeds.

The bark is astringent and used as a lotion for leprous ulcer. The powdered fruits are used for healing wounds. Locally, bark decoction is used to treat toothache and gum trouble.

Lavandula officinalis Chaix.

Family : Lamiaceae
Trade/Common Name : Lavender

A hardy evergreen fragrant flowering shrub upto 1.2 m tall with grey foliage. Flowers are highly esteemed for their fragrance in dried state.

It is native to Mediterranean regions. It is found in various parts of India. It is propagated by cuttings.

Flowers yield an essential oil (0.81%) by steam distillation. It conatins ester (50-53%) with an essence which contain a hydrocarban. It is used as stimulant, carminative, sternutatory and tonic. It is useful in hysteria, nervous headache and other nervous infections

Lavandula officinalis

Lavatera cashmerina Cambess.

Family : Malvaceae
Trade/Common Name : Wild hollyhock

A perennial herb upto 2 m tall. Leaves 3-5 lobed, lower leaves rounded heart-shaped with 5 shallow and toothed lobes. Flowers bright-pink with dark veined petals on terminal spikes.

It is found in forest clearings, meadows, shrubberies of western Himalaya from Kashmir to Uttaranchal. It is propagated by seeds.

It is a known medicinal herb in many Unani medicinal preparations. It is useful in throat problems and given as mild laxative.

Lawsonia inermis Linn.

Syn. *L. alba* L.

Family : Lythraceae
Trade/Common Name : Henna

A much branched shrub. Branches are usually 4-angled with sharp point ending. Leaves are opposite, acute and sharp pointed. Flowers are small, fragrant, appear in terminal large branches, white or pinkish in colour.

Lawsonia inermis

It is commercially cultivated as cash crop in Rajasthan, Punjab, Madhya Pradesh, Chhatisgarh and Gujrat. It is propagated by seeds or cuttings.

Leaves, bark and seeds are medicinally important. Leaves are astrigent. The decoction of leaves is used as gargle in sore throat. Paste of the leaves are applied for relieving headache and burning sensation in feet. It also applied on boils, burns and skin diseases. Leaves are used in the treatment of typhoid and haemorragia.

Leucas cephalotes (Roth) Spreng.

Family : Lamiaceae
Trade/Common Name : Dronpushpi

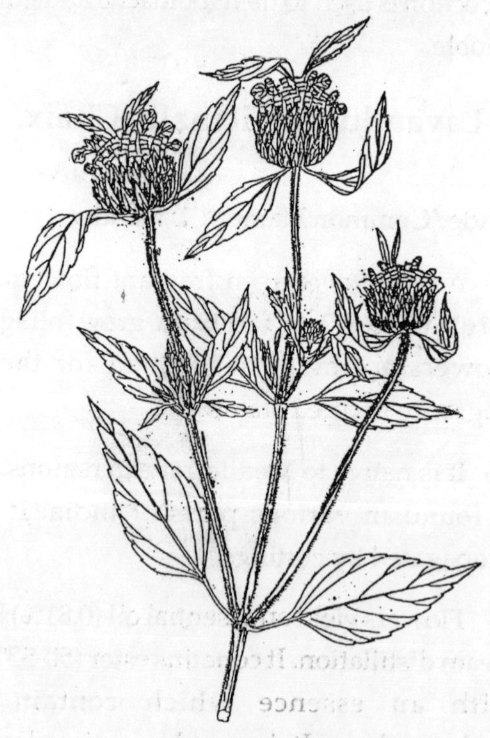

Leucas cephalotes

An erect, scabrous or pubescent, stout annual, 30-100 cm high. Leaves are shortly petioled, flowers small and white in colour, aggregated in axillary globose whorls and blooming almost throughout the year.

The plant is found as a weed throughout the greater part of India, ascending upto 1800 m in the Himalayas. It is propagated by seeds.

Seeds and flowers yield an essential oil. α-sitosterol and a glycoside which yielded glucose. It is considered to be a stimulant, diaphoretic, laxative, anathematic, antiseptic and insecticidal. A syrup of the flowers is used as a domestic remedy for cough and colds.

Leucas lanata Benth.

Family : Lamiaceae
Trade/Common Name : Dhurlughas

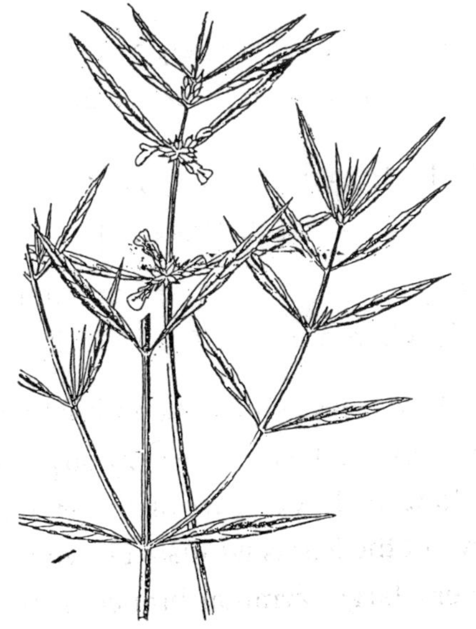

Leucas lanata

A tufted, woolly, tomentose, perennial herb with white flowers in axillary loose whorls.

It is found in grassy slopes, rocky boulders and dry field borders upto 2500 m of Himachal Pradesh. It is propagated by seeds.

The decoction of the leaves and apical shoots with gur is used as an abortifacient upto 3 months of pregnancy. Infusion of flowers are used in skin diseases. It is also used to treat piles.

Lilium candidum Linn.

Family : Liliaceae
Trade/Common Name : Lily

A hardy herbaceous plant attains 1.5 m height. Leaves are pale green and formed as soon as the flower stem has died down. Flowers are trumpet-shaped, glistening, white in colour with golden pollen. Flowers are 10 cm long, sweetly perfumed, 12-20 blooms appear on each stem.

It is native to Asia Minor. It is found in various parts of India. It is propagated by bulbs.

From the flower by extraction with petroleum ether 0.20 to 0.25% concrete is obtained which is waxy and green brown in colour. From the concrete 30 to 35% alcohol soluble absolute is yielded. The lily oil can be used in high grade perfumes.

Linum usitatissimum Linn.

Family : Linaceae
Trade/Common Name : Common Flax

An annual with fibrous stem herb upto 60 cm high. Flowers are delicate blue, mauve and white.

It is native to Europe, Eastern and Northern Africa. It is commercially cultivated in Madhya Pradesh, Chhatisgarh, Bengal, Jharkhand and Bihar. It is propagated by seeds.

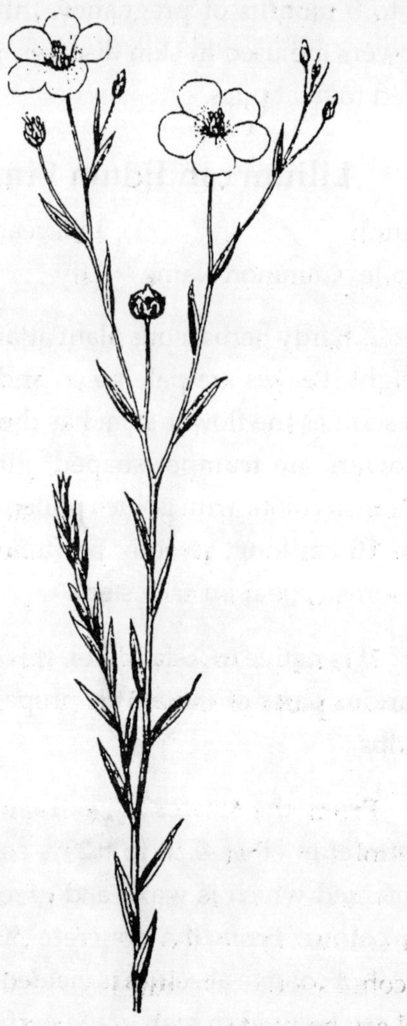

Linum usitatissimum

Seeds yield an unsaturated fatty oil and also hold a mucous. It is serve as demulcent and emollient. It is widely used as a sedative drink, for treatment in genito-urinary diseases.

Liparis paradoxa Rchb. f.

Family : Orchidaceae
Trade/Common Name : Wild Lily Orchid

An annual herb with tufted stem, base sometimes pseudobulbous. Leaves are 13 cm long and 2 cm broad. Scape stout, angular, to 25 cm long. Flowers are coriaceous upto 1 cm across, yellow brown in colour.

It is distributed in Uttaranchal, Naga and Khasia hills , Sikkim, Chhatisgarh and Karnataka. It is propagated by cuttings.

It is one of the eight drugs of the 'Ashtravargra' and used as a general tonic.

Lobelia nicotianaefolia Hyne. ex. Roth.

Family : Campanulaceae
Trade/Common Name : Wild Tobacco

An annual with stout and hallow stem upto 5 m tall. Leaves are very large upto 45 cm long in lower part of the stem, main nerve of the leaf is whitish. Flowers appear in very large terminal branches, white in colour.

Lobelia nicotianaefolia

It is found in Deccan peninsula. It is propagated by seeds.

It is used to treat asthma and bronchitis. It causes irritation of mucous membrane, toxic.

Lyonia ovalifolia (Wall.) Drude.

Family : Ericaceae
Trade/Common Name : Aiyaar

A hardy, deciduous tree, to 15 m high; bark reddish-brown, rough, peeling off in narrow strips. Leaves alternate or subopposite, ovate, ovate-lanceolate, apex acute or acuminate, base rounded. Flowers numerous, flask-shaped, arranged on 5-8 cm long racemes and blooming in the month of March-July.

It is native to Himalaya region and commonly found in Uttarkashi district of Uttaranchal between 500-2500 m elevation. It is also found in Khasi hills. It is propagated by seeds.

The extract of the leaves is rubbed on itching. The fusion of young leaves and buds are used in skin diseases.

Madhuca indica Gmel.

Syn. *Bassia latifolia* Roxb.

Family : Sapotaceae
Trade/Common Name : Madhua

A large deciduous tree upto 20 m tall with a speading habit. Leaves are clustered at the end of the branches. The fleshy, tubular, creamish white flowers with intoxicating smell appear at the ends of the branches during March-April. Flowers are stalked and drooping.

It is native to India, It is found in Chhindwara district, M. P. and also occur in all plains and lower hills of India and common in sub-mountainous regions of the Himalayas. It is propagated by seeds.

Leaves are astringent. The ash of leaves mixed with ghee is applied on burns and scalds. Flowers have cooling and nutritive and antibacterial properties and are used to treat bronchitis and cough. The bark prepartion is used for treating itch, ulcers and bleeding gums. Oils extracted from

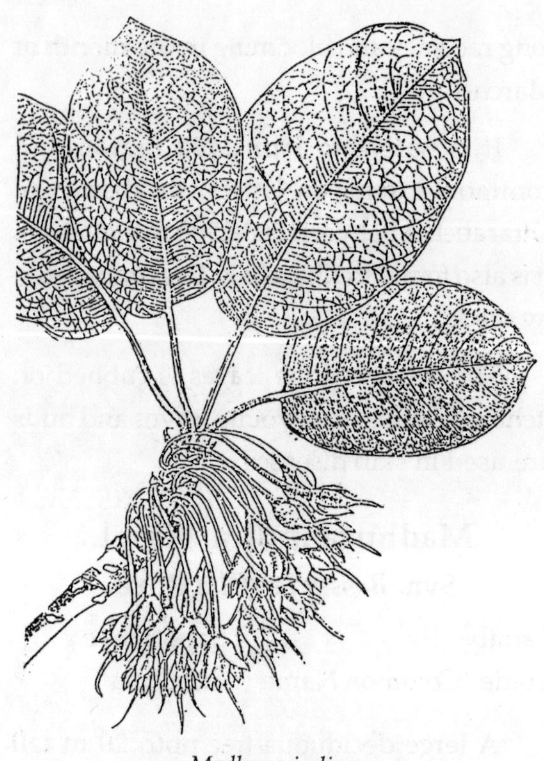

Madhuca indica

seeds are said to be laxative and used to treat skin diseases.

Malaxis cylindrostachya (L.) O. Kfze.

Family : Orchidaceae
Trade/Common Name : Bog Orchid

An orchid with stem like pseudobulb, clothed with leaf sheaths. Leaves are solitary, oblong or rounded to 10 cm long, blunt and plicate. Inflorescence densely many flowered upto 45 cm long. Flowers are pale yellowish green, short leaved.

It is distributed in Kashmir and Sikkim. It is propagated by bulbs.

The pseudobulb preparation is considered as tonic.

Mallotus philippinensis (Lam.) Muell. Arg.

Family : Euphorbaceae
Trade/Common Name : Kamila

A much branched, evergreen tree, 6-9 m high, with a trunk of 90-120 cm round. Young branches and inflorescence are rusty-pubescent. Leaves are alternate, long stalked. Flowers are small, in panicled racemes and blooming in the month of October to continue upto February.

Fruit contains resin (80%), tannin acid, gum and volatile oil. The most important constituent is a brownish resin composed of crystalline substance called rottlerin. It is anthelmintic, antioxidant, aphrodisiac, blood purifier, purgative in full doses, occasionally causing nausea, but seldom vomiting. Kamala is also used in parasitic

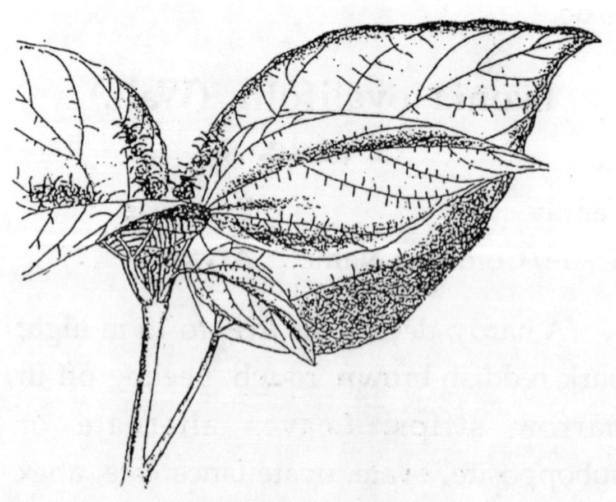

Mallotus philippinensis

affections of the skin, particularly scabies, pimples and ringworm, weeping, eczema, boils and also useful in bile trouble.

Mangifera indica Linn.

Family : Anacardiaceae
Trade/Common Name : Mango

A large, evergreen tree with a dense rouded or globular crown. Leaves are simple, petiolate, alternate, irregularly placed along the brnchlets. The fruit is a large drupe derived from one carpel of the flower.

It is native to India, it is found different parts of India and commercially cultivated in different states of the country. It is propagated by seeds or grafting.

The fruit is arich source of vitamin A and C. The bark is astrigent. It is used in diptheria and rheumatism. It possesses a tonic action on the mucous membrane. The rind is astrigent, stimulating tonic in debility of stomach.

Manikara archras Fosberg.

Family : Sapotaceae
Trade/Common Name : Sapota

A slow growing evergreen tree upto 9 m tall. Leaves are elliptic to ovate. Flowers are solitary in leaf axils.

Mangifera indica

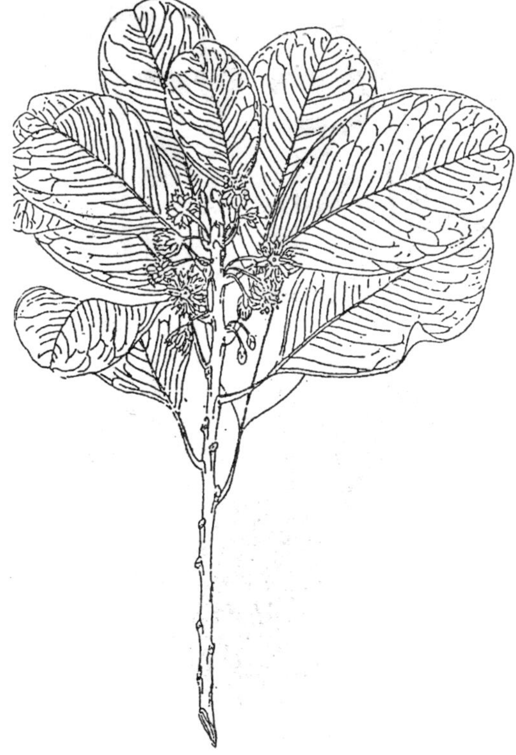

Manikara archras

It is native to Mexico. It is cultivated in different states of India. It is propagated by seeds, inarching, air-layering and also by cutting.

Bark have several medicinal uses. The latex is used for making chewing gum.

Marrubium vulgare Linn.

Family : Lamiaceae
Trade/Common Name : White Horehound

A perennial strong smelling herb. Stem branched, densely woolly-haired and forming clumps. Leaves rounded wrinked, toothached and densely woolly-haired. Flowers are small arranged in axillary whorls, whitish 2-lipped.

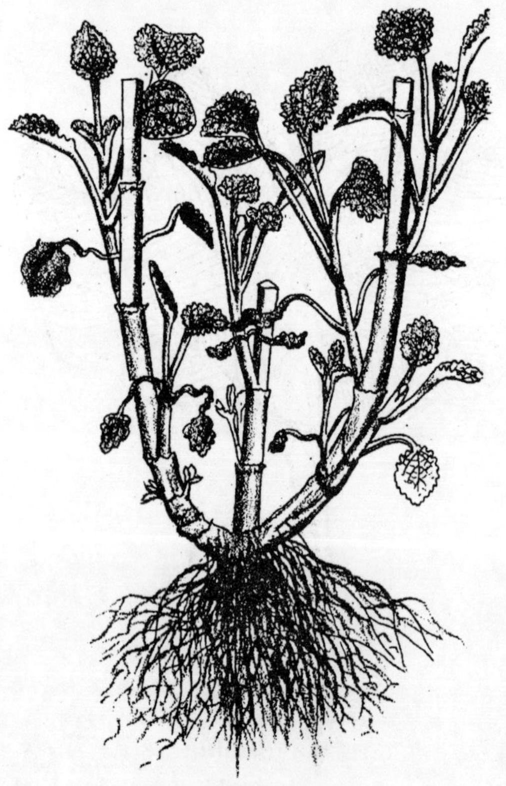

Marrubium vulgare

It is found in wastelands, roadsides, foothills, dry pastures in the western Himalayan regions. It is propagated by seeds.

Dried flowering tops contain a bitter principle, marrubin (0.4%), tannins, an essential oil and mucilage. It possesses bitter, tonic, expectorant, diuretic, carminative and laxative properties. It is useful in cold and cough. As a home remedy, its extract has been mixed in cough mixtures, lozenges, intravenous administrations produces hypotensive effects.

Marsilea minuta Linn.

Family : Marsileaceae
Trade/Common Name : Marsile

A fern with four lobes petiolated leaves.

It is found at Alwar, Ajmer, Mt. Abu and Sirohi of Rajasthan. It is propagated by runners.

The decoction of leaves along with ginger is used in cough and bronchitis.

Matricaria chamomilla Linn.

Family : Asteraceae
Trade/Common Name : Chamomile

A perennial, stongly aromatic, variable herb upto 1 m tall. Leaves much divided with linear segments. Flower heads white carries on long penduncles.

It is found in waste places, roadsides, fields, orchards and cultivated crop fields in upper gangatic plains, Punjab, Kashmir. Presently, cultivated as cutflower. It is propagated by seeds.

Flower yields an anthemic acid, azulene, chamomallin, alpha-bisabolol and cyclic-ether. It possesses stimulant, attenuant and carminative properties and used in constitutional debility, hysteria, dyspepsia and intermittent fevers. Its oil is used in rheumatism, effective in flatulence and colic. Sometimes its extracts used in hair dyes, mouth washes and shampoos.

Matricaria chamomilla

Mecanopsis aculeate Royle.

Family : Papaveraceae

Trade/Common Name : Wild Yellow Poppy

A biennial herb having sparsely bristly stem and leaves and light blue papery flowers with prominent yellow stamens. Leaves deeply and irregularly pinnately lobed.

It is found in rocky slopes, shady and moist places in western Himalaya from Kashmir to Uttaranchal. It is propagated by seeds.

It is used as afolk medicine. Water extract of whole herb including flowers used to wash wounds.

Mecanopsis aculeate

Melastoma malabathricum Linn.

Family : Melastomaceae
Trade/Common Name : Malabar Gooseberry

A shrub in which mauve-purple flowers are produced in terminal, corymbose panicles.

It is distributed in Indomalayan region, occurring at Junapani forest of Chhindwara district of M.P. and also in other parts of India in humid places.

The decoction of flower (about 20 ml) is taken in the morning for one month in the spermatorrhoea. The leaf juice if taken in diarrhoea and dysentry.

Melia azadarach Linn.

Family : Meliaceae
Trade/Common Name : Bead Tree

A moderate-sized deciduous tree, 9-12 m high with a cylindrical bole. Bark dark grey with shallow longitudinal furrows; leaves bi- or occasionally tri pinnate; leaflets ovate or lanceolate, serrate. Flowers lilac, fragrant, in axillary panicles.

It is native to West Asia. It is found wild in the sub-Himalayan tract up to 1,800 m altitude. Sometimes, it is cultivated in the plains as an ornamental avenue tree. It is propagated by seeds or cuttings.

Leaves, fruits and seeds are useful in skin diseases as well as in rheumatic pains.

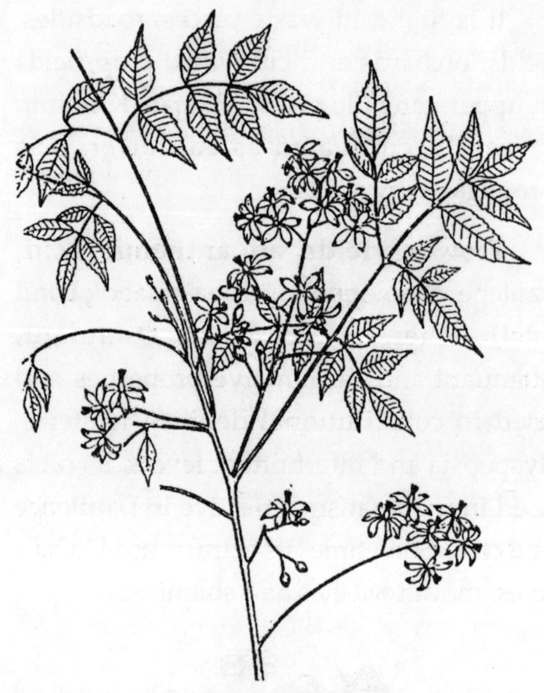

Melia azadarach

Fruit and leaves as antiseptic and wormicides. Fruit pulp regarded as an abortifacient.

Melissa parviflora Benth.

Family : Lamiaceae
Trade/Common Name : Bililotan

A hardy herbaceous perennial plant of 1 m tall. Leaves are lemon scented. Flowers are white or yellow.

It is native to Mediterranean region. It occurs in temperate Himalayan region from Uttaranchal to Sikkim and Mishmi and Khasia hills. It is propagated by seeds or division of roots.

It is stimulating and anti-spasmodic. The tea prepared from these plants is used

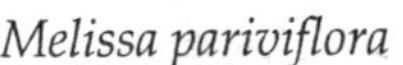

Melissa pariviflora

Mentha arvensis

for flatulence, lack of appetite and indigestion. It is also useful in nervous complaints, hysteria, cardialgia, spasms, vertigo, nigra and in all cases of general atony.

Mentha arvensis Linn.

Family : Lamiaceae
Trade/Common Name : Japanese Mint

An erect branched perennial herb with runing rootstocks and rigid stem upto 75 cm tall. Leaves are 5 cm long, leaf stalk small or none, margins toothed. Flowers are lilac in colour.

It is cultivated in tarai and Haldwani area in Uttaranchal, Jammu & Srinagar and some parts of Punjab. It is propagated by runners.

Leaves yield an essential oil (5%), which contains menthol (40-50%) and dementholised oil (50-60%). It is antiseptic, carminative, refrigerant, stimulant and diuretic. The plant is used to treat liver and spleen diseases, asthma and jaundice.

Menthol is used in medicine for stomach disorders and in ointments for headache. The infusion of leaves is used in indigestion and rheumatic pains.

Mentha longifolia (L.) Hunds.

Syn. *M. sylvestris* L.

Family : Lamiaceae
Trade/Common Name : Horse-mint

An erect or diffuse, aromatic, tomentose herb with white pubescent, lanceolate and toothed leaves. The flowers are pink in slender terminal spikes and blooming in the month of June to continue upto August.

The herb is found in the temperate Himalayas and Western Tibet at the altitudes of 1200-3600 m. The Kashmir, Himachal Pradesh, Uttaranchal, it is found growing in abundance in along streamsides, waysides, bogs, irrigation channels, borders of paddy fields. It is propagated by seeds.

Leaves and top-shoot yield essential oil, contains piperitone oxide, diosphenol, piperitenone. The leaves are astringent and are used for rheumatic pains.A decoction of the plant is used in fever and heat apoplexy. It is much used as an incense, as masala in washing the hair and is useful in the stomach disorders, dyspepsia, vomiting and is case of the pain in the liver or womb. The plant is stated to be useful in scorpion sting and snake bites.

Mentha piperita Linn.

Family : Lamiaceae
Trade/Common Name : Pepper-mint

An erect, hardy, glomous, branched perennial herb with runing rootstocks upto 90 cm tall. Leaves are long, petiolated, opposite, lanceolate.

Mentha piperita

It is cultivated in some parts of Tarai regions of India. It is propagated by runners.

Flowering tops contain essential oil (03-07%). It is used as carminative, antiseptic, gastrostimulent and flavouring agents. It is used in lotion and fermentations externally as a resolvent for bruises and scabies. The essence is applied for toothache and swollen gum. It is administered in serious nervous disorders, in headache, colis dischrges with nervous symptoms. As a flavouring agent, peppermint is used in wide range of pharmaceuticals, confectionery, alcohol drink, dental creams and mouth washes.

Mentha spicata Linn.

Syn. *M. viridis* Linn.

Family : Lamiaceae
Trade/Common Name : Spear-mint

A perennial her upto 1 m tall. Flowers are borne in long spikes, white or pink in colour.

It is native to North America. It is cultivated in some parts of India. It is propagated by runners ot cuttings.

Mentha spicata

Flowering tops yeild an essential oil (0.25-0.5%). The essential oil contains carvone (55.8%). It is carminative, stimulant and to prevent vomiting. Tea made from this herb is used to treat neuralgia, indigestion and diarrhoea.

Menyanthes trifoliate Linn.

Family : Menyanthaceae
Trade/Common Name : Buckbean

A perennial herb with characteristic long-petiolated purplish trifoliate leaves. Flowers are pinkish white in colour and arranged in terminal compact spikes.

Menyanthes trifoliate

It is found in swampy meadow, marshes, peat bogs, ditches, edges of lakes and ponds from South to North Himalaya. It is propagated by seeds.

It contains glycoside-menyanthin, meliatin and tannins. It is used in the intermittent, and remittent fever, gout, herpetic complaints, rheumatism, dropsy, scurvy and worms.

Mesua ferrea Linn.

Family : Guttiferae
Trade/Common Name : Nagkesur

An erect herbaceous with white flower.

It is occurring in Eastern Himalayas, Assam, West Bengal, Western Ghats and Andaman Islands. It is propagated by cuttings.

Mesua ferrea

The seeds yield an oil. It is used to treat rheumatism and skin diseases. Dried flowers are used for preparation of perfumed ointments.

Michelia champaca Linn.

Family : Magnoliaceae
Trade/Common Name : Champaka

An evergreen tree, to 25 m high; bark dark-grey; branches densely tomentose; buds grayish brown. Leaves ovate-lanceolate to elliptic; lateral nerves 12-18 pairs, arched; stipules oblong-acuminate, silky outside. Flowers solitary, axillary, pale-yellow and blooming in the month of April-June.

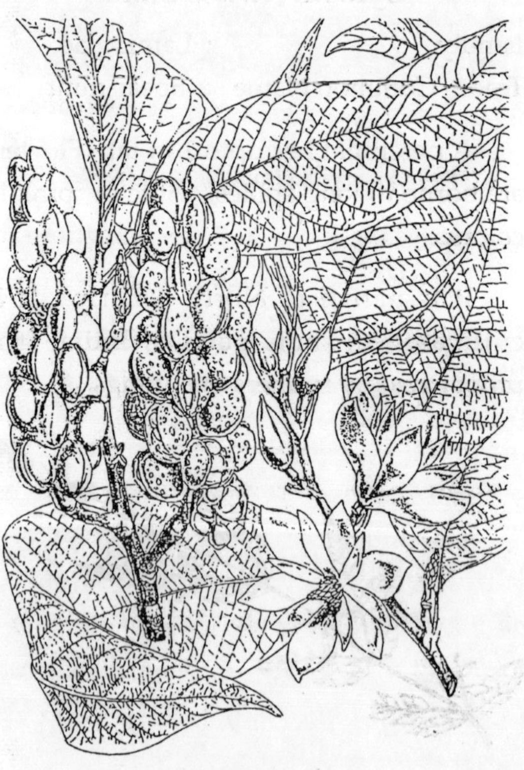

Michelia champaca

It is native to E. Himalyas, Assam and Western Ghats. It is cultivated throughout India. It is propagated by seeds.

Bark in used as diuretic, febrifuge and stimulant. The juice is used to treat colic. The oil extracted from flowers are employed in perfumery industry. It is also cephalagia, opthalmia, gout and rheumatism. Flowers and fruits are also used as stimulant, diuretic, stomachic and antispasmodic.

Microstylis muscifera (L.) Ridl.

Family : Orchidaceae
Trade/Common Name : Adder's Mouth Orchid

A terrestial orchid with tuberous and sheathed stem upto 40 cm tall. Leaves are 10 cm long and rounded. Inflorescence upto 25 cm long. Flowers are 4 mm in diameter, yellowish green in colour

It occurs in temperate Himalayas from Kashmir to Sikkim between 2600-3300 m elevation. It is propagated by seeds or cuttings.

Roots are regarded as tonic and lactagouage.

Mimosa pudica Linn.

Family : Mimosaceae
Trade/Common Name : Sensitive plant

A diffuse undershrub, 50-90 cm high. Stem and rachis clothed with prickles; leaves bipinnate; pinnate 2-4, digitatively arranged, with 10-20 pairs of leaflets. Flowers in pinkish globose heads.

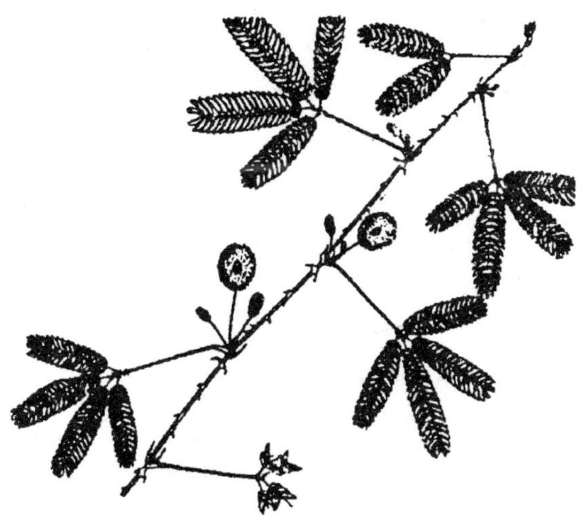

Mimosa pudica

It is native to America but naturalized throughout India. It is propagated by seeds or cuttings.

The plant contains beta-amyrin, beta-sitosterol and friedelin. Mucilage of seeds contain galactose and mannose. An adrenaline-like substance identified in extracts of leaves. It is astringent, cooling, antiseptic, alternative and blood purifier. It is resolvent, alternative and carminative. It is used in buring sensation of body, diarrhoea, dysentery, haemophilic condition, leucorrhoea, morbid condition of vagina. Root is aphrodisiac, cooling, vulnerary; cures kapha, asthma, biliousness, vaginal and uterine complaints. Juice is antiseptic, alternative and blood purifier. Leaves rubbed into paste applied to hydrocele. Leaf and stem in scorpion sting.

Mimusops elengi Linn.

Family : Sapotaceae
Trade/Common Name : Indian Medlar Tree

An evergreen, glabrous tree, to 12 m tall' bark dark-grey. Leaves elliptic-ovate to lanceolate, base acute or attenuate, apex acute or acuminate, entire, subcoriaceous; petioles to 3 cm long. Flowers brownish-white, scented, to 1 cm across and blooming in the month of March-June.

It is native to Peninsular India. It occurs in various parts of western Ghat. It is propgated by seeds.

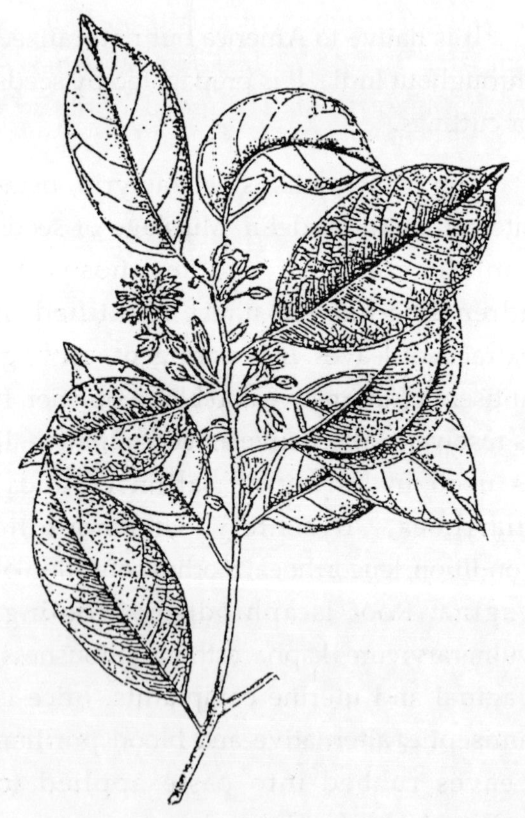

Mimusops elengi

Bark is used to treat diarrhoea and dysentery. A decoction of bark is used as gargle. The fruit and bark possess tonic and astrigent properties. Ethno-medicinally, powdered seeds are pasted with oil and used as suppositories in case of obstinate constipation especially children.

Mirabilis multiflora (Torr.) Gray.

Family : Nyctaginaceae
Trade/Common Name : Wild Four O'Clock

A hardy tuberous rooted perennial much branched herb upto 1 m tall. Flowers are purple, around 5 cm long.

It occurs in the foothills of Himalayas between 1000-3000 m elevation. It is propagated by seeds or divisions of tubers.

Root is used to treat stomachic.

Momordica charantia Linn.

Family : Cucurbitaceae
Trade/Common Name : Bitter Gourd

A annual, slender, climbing or trailing herb. Leaves are pubescent, suborbicular. Flowers monoceious, solitary and yellow.

It is cultivated in various parts of India and widely distributed throughout the country. It is propagated byseeds.

Seed yield an essential oil, Fruit contain bitter glycosides, cucurbitacins. The fruit is used as stomachic, carminative. Leaves are

Momordica charantia

purgative, emetic and bitter properties. The fruit is uses for the treatment of rheumatism, gout, diseases of liver and spleen and useful in diabities. It is also used in leprosy, piles and jaundice.

Morchella esculenta (L.) Pers.

Family : Helvellaceae
Trade/Common Name : Honey Combed Mushroom

Guchhi is an ascomycetes having pale-white stalk which supports the thick, erect, fruiting body, bearing a rounded or conical pileus covered with a number of prominent ridges. It is yellowish brown to olive coloured, often dark brown also. It appearing in the decaying organic matter in forests during March-August.

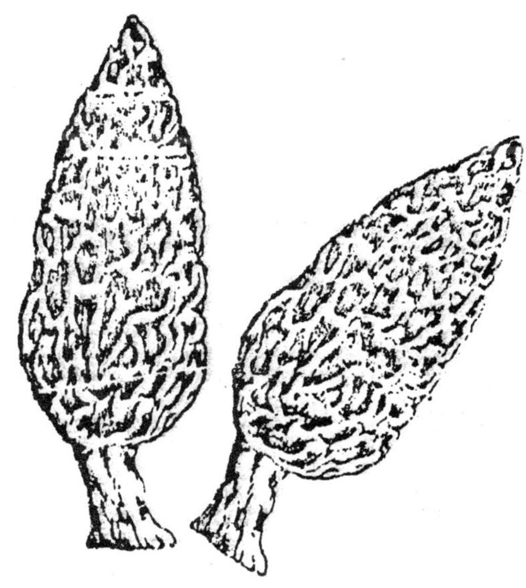

Morchella esculenta

It is a valuable mushroom growing naturally in humus rich forest soils in Uttaranchal and Himachal Pradesh during march-April between 2200-3200 m. It is propagated by span.

It contains moisture (89%), protein (3.6%), carbohydrate (5.73%), fat (0.27%) and ash (1.34%). It is eaten both fresh and dry as vegetable and as rich man's food. It is also considered as tonic, aphrodisiac and narcotic. As a domestic remedy, the paste of the mushroom is applied to cure boils that have become chronic.

Morina coulteriana Royle.

Family : Dipsacaceae
Trade/Common Name : Bright Pop'

A perennial herb upto 1 m tall with a basal cluster of spiny-lobed thistle-like leaves. Stem leaves linear-oblong, stalkless with spiny margins. Flowers yellow in long terminal, interrupted spikes. Bracts leaf-like and longer than the flower whorl.

It is found open slopes, leftover forests, dry situations altitude above 2500 m. It is propagated by seeds.

The roots used as incense.

Moringa oleifera Lamk.

Family : Moringaceae
Trade/Common Name : Drumstick Tree

A small tree, 9-12 m tall with corky bark. Wood is soft and taste pungent. Leaves are 3-pinnate and leaflets are in 6-9 pairs with an odd leaflet elliptic-ovate or obovate. Flowers are pedicelled, white and honey scented in spreading, paniculate racemes and blooming in the month of January to continue upto April.

The tree is found in the outer Himalayas at low altitude and is cultivated throughout the hotter parts of India and tropical countries. It is propagated by seeds or cutting.

Root and its bark contain spirochic, pterygospermin (an antibiotic), moringine and moringinene. The plant is used as a tonic, abortifacient, stimulant, aphrodisiac and rheumatism. Oil obtained from the seeds is a useful remedy for gout and acute rheumatism. Leaves, bark and roots are considered spasmodiac. Bark offers a cure for gum boils and toothache. Sporochin enhances the growth of epithelial cells and suppresses the irregular contraction of the uterus and provides strength to it.

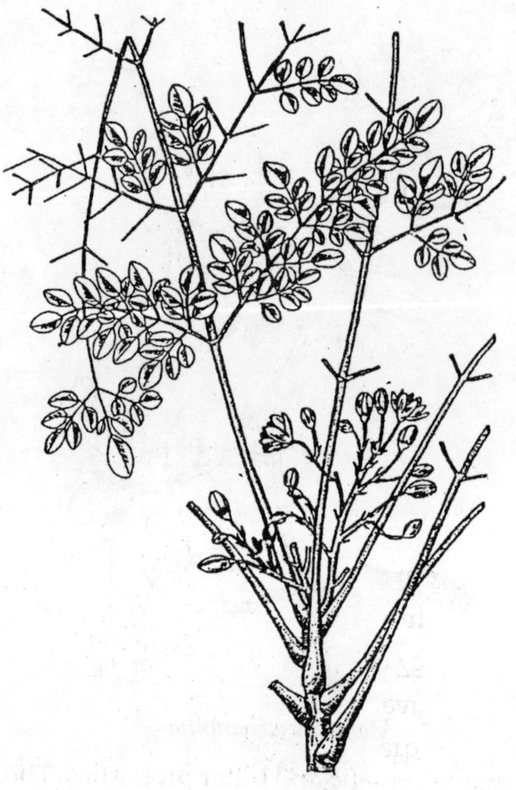

Moringa oleifera

Morus alba Linn.

Family : Moraceae
Trade/Common Name : White Mulberry

A tall tree with trunk upto 60 cm in diameter. Leaves are irregularly, toothed, pale green in colour. Fruits are berries of 5 cm long, sweet, white to purple.

It is native to China, Europe and North America. It occurs throughout the India and also cultivated as a selter of silkworm. It is propagated by seeds, cuttings or grafting.

Morus alba

Ripe fruits and roots are medicinally important. A tea made from root is used to treat diarrhoea. Fruit preparation is given for high fever and also used as laxative.

Mucuna pruriens (L.) DC.

Syn. *M. prurita* Hook.

Family : Papilionaceae
Trade/Common Name : Cowhage

An annual, half-woody, twining herb with long, cyclindrical branches. Leaves are alternate, 3-foliate; petiolate long, ovate or rhomboid. Flowers are solitary or 2-3 together bearing purplish colour, arranged raceme and blooming in the month of July to continue upto September.

The herb is found almost all over India and foothills of Himalayan regions of Uttaranchal and Himachal Pradesh. It is propagated by seeds.

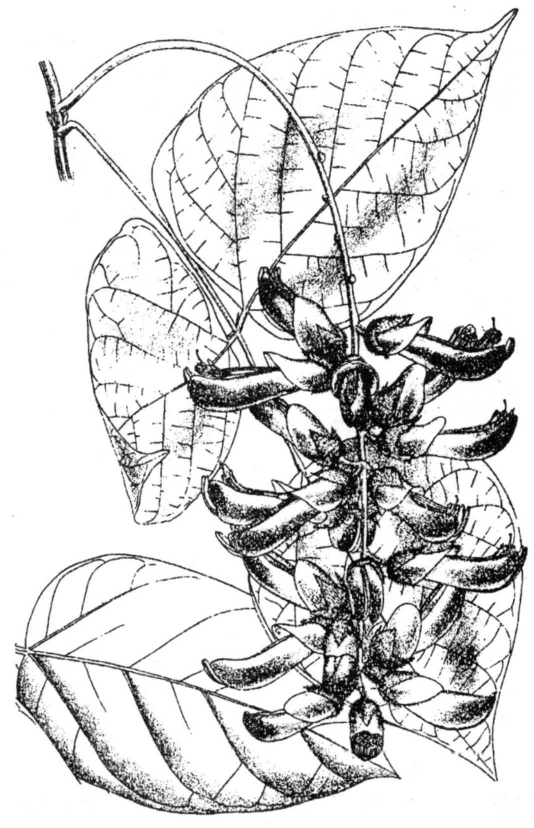

Mucuna pruriens

The seeds contain alkaloids mucuadine, mucunadinine, Prurienidine alongwith mucunin, mucunadine and small amount of nicotine. Cowhage is anthelmintic (hairs), aphrodisiac, astringent, diuretic and nervine tonic. It is useful in cholera (in the form of a strong infusion), dropsy, general debility, leucorrhoea, seminal weakness and as a tonic.

Murraya koenigii Spreng.

Family : Rutaceae
Trade/Common Name : Curry leaf

A handsome, aromatic, more or less deciduous shrub or a small tree, upto 6 m in

Murraya koenigii

height and 15-40 cm in diameter. Bark is dark brown in clolour and the leaves are imparipinnate. Flowers are white in colour , fragrant and arising in terminal, corymbose panicles and blooming in the month of Aril-May.

The plant is found, almost, throughout India, upto an altitude of 1500 m. It is propagated by seeds.

Leaves yield an essential oil and glucoside-koeinigin. It is used in the preparation of a tonic. Bark and roots are used as stimulants and applied externally to cure eruptions and bites of poisonous animals. As decoction with bitters, it acts as a febrifuge and is also used in snake bite. Leaves are frequently used in curries.

Murraya paniculata (Linn.) Jack.

Family : Rutaceae

Trade/Common Name : Marchula

An evergreen, large shrub or small tree, to 5 m high; stem woody, rough, pale-white, twigs pubescent. Leaves imparipinnate, 4-15 cm long; leaflets 3-7 alternate or subopposite, ovate-lanceolate, shortly stalked, tip acuminate, shining on both surfaces, gland dotted and hairy only along mid rib beneath. Flowers white, fragrant, in terminal or axillary corymbs and blooming in the month of March-April.

It is commonly found in the shady ravine slopes of upto 1500 m elevation in Uttaranchal Himalayas. It is propagated by seeds or cuttings.

Leaves yield an essential oil (0.01%). It possess stimilant, tonic, stomachic, astringent and refrigerant properties. The leaves are used to treat cough; rheumatism and hysteria.

Musa paradisiaca Linn.

Family : Musaceae

Trade/Common Name : Jangli Kele

A monceious perennial herb with rhizome and false aerial stem upto 10 m tall. The infloreseance arises from rhizome and appear at the top of the aerial stem.

It is native to Tropical Asia. It occurs in some parts of Chhindwara district of

Musa paradisiaca

Madhya Pradesh. It is propagated by suckers.

Ethno-medicinally, root and stem decoction (about 20 ml) are taken in the morning for one month as tonic in blood and veneral diseases.

Myrica esculenta Buch. Ham.
Syn. *M. nagi* Hook . f. (non Thunb)

Family : Myricaceae
Trade/Common Name : Box-myrtle

A small moderate-sized evergreen tree, 3-15 m high. The bark is gray or brownish gray and the leaves are lanceolate. Flowers are minute and blooming in the month of August to continue upto December.

Bark is rich in tannins and contains the glycoside. Fruits are edible and are used in the preparation of refreshing drink. These are considered pectoral, sedative, stomachic and carminative. The bark of the plant is astringent, carminative and antiseptic. A decoction of the bark is useful in asthma, diarrhea, fevers, lung affections, chronic, bronchitis, dysentery and diuresis. The bark is chewed to relieve toothache. It is indicated in cough, gum trouble, earache, gonorrhoea, piles and throat trouble.

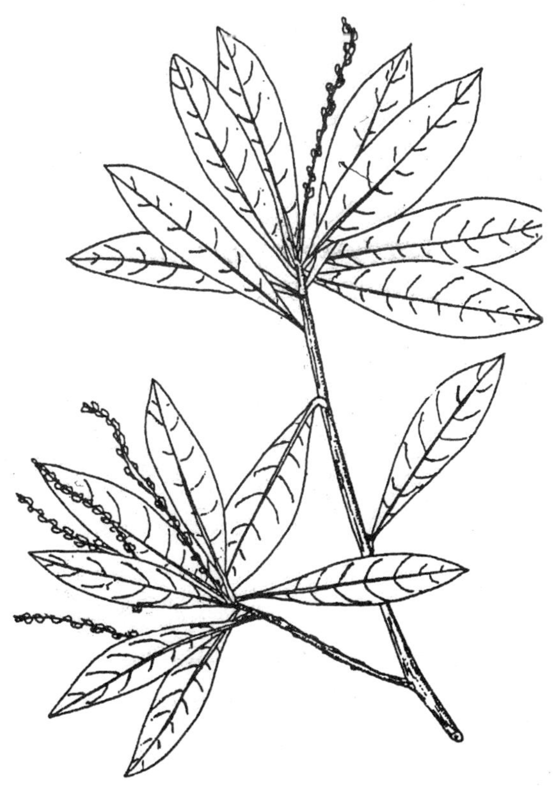

Myrica esculenta

Myristica fragrans Houtt.

Family : Myristicaceae
Trade/Common Name : Nutmeg

A evergreen tree with leafly undivided trunk of 10 m tall. Leaves are shining, deep green, alternate, oblong, entire and aromatic. Flowers are white or pale yellow, small and fragrant, borne in cluster from the axils of leaves. Fruits are pear-shaped, golden yellow in colour. Seeds are covered by red fleshy part, known as aril which are dried and sold as mace.

It is native of East Moluccas. It is cultivated in Tamil Nadu, Kerala and Assam. It is propagated by seeds or cuttings.

The dried ripe seed and mace, the dried arillode is steam-distilled to obtain an essential oil (6.5-8.0%). It contains camphene, pinene, dipentene, myristicin, eugenol, aldehyde, etc. The oil of nutmeg is used in perfumes and for flavouring dentifrices, baked foods, cakes, cookies, puddings pickles, sauces and savoury preparations. For treatment of inflammations of bladder and urinary tract, the oil is utilized in small doses in pharmaceutical preparations.

Myristica fragrans

Myroxylon balsamum (L.) Harms.

Family : Fabaceae
Trade/Common Name : Tolu balsam

An attractive, evergreen tree upto 30 m tall. Bark is rough when injured a yellow-brown resin in excreted. Flowers are yellow in colour.

It is native Argentina and Brazil. It is also successfully grown in India. It is propagated by seeds or cuttings.

It contain resinous material (75-80%) like cinnamic and benzoic acid. It is an ingredient of tincture benzoin. Resin is used in lozenges for the relief of throat irritation and coughs. For the treatment of respiratory ailments it is heated and vapour inhaled.

Myrsine africana Linn.

Family : Myrsinaceae
Trade/Common Name : Baibidang

An erect shrub, 0.6-1.2 m high. Leaves sharply toothed; flowers minute, white in colour and blooming in the month of April to continue upto August.

Myrsine africana

It is found in the outer Himalayas from Kashmir to Nepal and Khasi hills at altitudes of 300-2700 m. It is propagated by seeds.

Fruit contains embelin (3%), quercitol, emobellin, etc. It is edible and has anthelmintic properties and effective for expulsion of the tape worm. It is also as a dropsy and colic. Decoction of the leaves is used as a 'blood purifier'.

Nardostachys grandiflora DC

Syn. *N. jatamansi* DC.

Family : Valerianaceae
Trade/Common Name : Spikenard.

An aromatic, erect, perennial herb, 10-60 cm high with long stout, woody rhizomatic rootstock covered with tail brown fibres left over from the withered leaves towards the stem, while the root continues to penetrate deep in the soil. Leaves are radical in nature long, narrow and the flowers are creamy white, often rosy or pale pink in appearance arising in terminal corymbose cymes and blooming in the month of June-July.

It is found in the alpine Himalayas from Himachal Pradesh to Sikkim and Bhutan, at altitudes between 3000-5000 m. It is propagated by root cutting.

Nardostachys grandiflora

Roots contain essential oil having jatamansone, jatamansic acid, lupeol, jatamansinone, jatamansinol, jatamansin, angelic acid valeranone, valeranal. Rhizome also contains alkaloid-actinidin. It is considered as tonic, stimulant, anti-spasmodic, diuretic, deobstruent, emmengogue, stomachic and laxative. An infusion of the rhizome is reported to be useful in epilepsy, hysteria, leprosy, respiratory disorders, nervous excitement, palpitation of the heart and a tincture is given in intestinal colic and flatulence.

Nelumbo nucifera Gaertn.

Syn. *Nelumblium speciosum* Willd.

Family : Nymphaeceae
Trade/Common Name : Indian lotus

A perennial with thick underground stoniferous rhizomes. Leaves circular upto 1 m diameter, peltate, stalk 1-2 m long carrying the leaf to water level for floating. Flowers white or pink, showy, with many whorls of petals.

It is found in lakes, ponds in throughout India. It is propagated by June to continue upto August.

Flowers cooling, used as astrigent in diarhoea, also cholera, in fever and diseases of liver, recommended as a cardio-tonic. Seeds used to check vomiting, diuretic, form a cooling medicine for skin diseases and leprosy. Locally, root powder is used for piles, also for dysentery, used as a paste in skin affections and in ringworm. Flowers are used for decorative and religious purposes and are consisdered sacred by Hindus.

Nepeta cataria Linn.

Family : Lamiaceae
Trade/Common Name : Catmint

A hardy, erect, perennial herb upto 1 m tall. Leaves are more or less heart shaped, opposite, coarsely dentate upto 75 cm long. Flowers are tubular, small, appear in clusters, whitish to pale purple in colour.

It is native to Europe and Temperate Western Himalayas. It is propagated by seeds or divisions of roots.

The plant is slightly acrid and bitter in taste and used as aperative and sudorific. Dried leaves and stalks are smoked to relieve respiratory ailments. A tea made from the plant is used to treat stomach ailment, nervous conditions and treating cold. Its

Nepeta cataria

poultice is used to reduce swellings. It is also used to delay menstruation. Locally, Leaves are chewed to relieve toothache.

Nepeta elliptica Royle ex. Benth.

Family : Lamiaceae
Trade/Common Name : Zoofa

An annual, erect herb with purple flower. Flowers are gynodioceous.

It is native to Tempertae Eurasia. It occurs in the alpine zone between 2700 to 3000 m altitude at Bhagrathi Valley, Herkeedun and Bhilagana Valley of Uttaranchal. It is propagated byseeds.

Nepeta elliptica

Its seed's infusion are used as carminative and useful to treat dysentery and uterine disorders.

Nepeta graciliflora Benth.

Family : Lamiaceae
Trade/Common Name : Upryaghas

An erect, much branched, slender herb or undershrub, to 50-150 cm high; stem stellate, pubescent. Leaves ovate-cordate, crenate-serrate, acute; petioles long. Flowers lilac or pink, stalked, in lax, many-flowered panicles and blooming in the month of February-April.

It is native to Himalaya and occurs in Gamari of Uttarkashi in Uttaranchal at 2500 m elevation. It is propagated by seeds.

The leaf paste is applied on forehead in headache. The paste is also applied in swelling due to injury.

Nepeta hindostana (Roth.) Haines.

Family : Lamiaceae
Trade/Common Name : Indian Catmint

A decumbent-ascending or erect herbs, 20-60 cm high; stem branched from the base, terete or 4-angular, densely hairy. Leaves broadly ovate or orbicular, rounded or cordate base, crenate-dentate, subobtuse, adpressed hairy. Flowers blue-purple in colour, in dense, usually peduncle cymes (whorls), forming interrupted slender

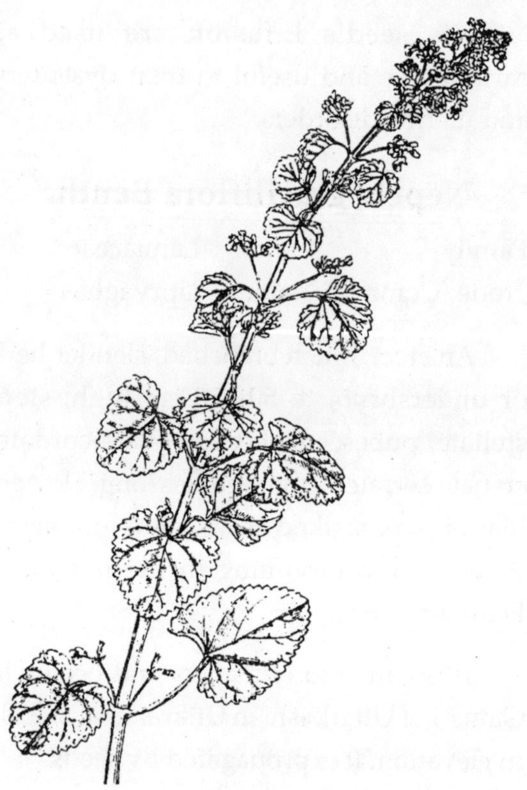

Nepeta hindostana

panicles and blooming in the month of January-July.

It occurs sporadiaclly on the moist slopes of foot hills upto 1500 m in Uttaranchal It is propagated by seeds.

Leaves are used as carminative, refrigerent and stimulant properties. The decoction of the plant is used in sore throats. The plant is administered as cardiac tonic and also in fever.

Nerium indicum Mill.

Family : Apocynaceae
Trade/Common Name : Kaner

A large, evergreen shrub, with milky juice. Leaves are mostly in whorls of 3. Flowers red or white in colour fragrant in terminal cymes and blooming in the month of April to continue upto June.

It is found in the Himalayas from Nepal Wastwards to Kashmir upto 1900 m, in Gangatic plain and Madhya Pradesh. It is propagated by seeds.

Root, bark and seeds contain cardioactive glycosides formerly designated as neriodorin, nerioderin and karabin. The bark also contains scopoletin and scopoline. The leaves yield oleandrin, nariodin, ursolic acid, etc. It is regarded poisonous; is reported as anthelmintic, cardiac, carminative, diaphoretic, febrifuge,

Nerium indicum

ophthalmic, powerful heart poison and repellant. Its local application is useful in swelling, leprosy, skin diseases and ulcers.

Nicotiana tabacum Linn.

Family : Solanaceae
Trade/Common Name : Wild Tobacco

A half hardy, annual, erect, viscid-glandular, much branched herb, to 1.2 m high. Lower leaves large, oblong or ovate-lanceolate, upper leaves sessile, smaller, elliptic or obovate, base semiamplexicaule, subentire, sinuate to repand, glabrous or hairy, acute or acuminate. Flowers reddish or pink-white, in cymose panicles and blooming in the month of November-January.

Nicotiana tabacum

It is cultivated as annual crop in various parts of the country. It is propagated by seeds.

Leaves are applied to skin to cure scabies, destroy lica and have a salutary effect is cases of neuralgic pains and toothache. As a rectal injection, it can acts as a direct stimulant in obstinate constipation and helps to reduce hernia or as a revulsive in apoplexy and asphyxia.

Nyctanthes arbor-tristis Linn.

Family : Oleaceae
Trade/Common Name : Parijata

A hardy, large shrub or a small tree upto 10 m high, with gray or greenish white rough bark and quadrangular branches. Leaves are opposite, ovate, acute, coriaceous covered with stiff white hairs. Flowers are small in size, sweet scented, sessile, 3-7 together in pedunculate heads, arranged in trichotomous cymes. Flowers tubular, orange red with white limb and blooming in the month of July to continue upto October.

The plant is a native of India, occurring wild in the Sub-Himalayan region from Chenab to Nepal upto 1500 m at altitude.

Flower yield an essential oil myctanthin (identical with α-crocetin from saffron), tannin, α-manital and glucose. The leaves contain α-amyrin, α-sitosterol, glycosides,

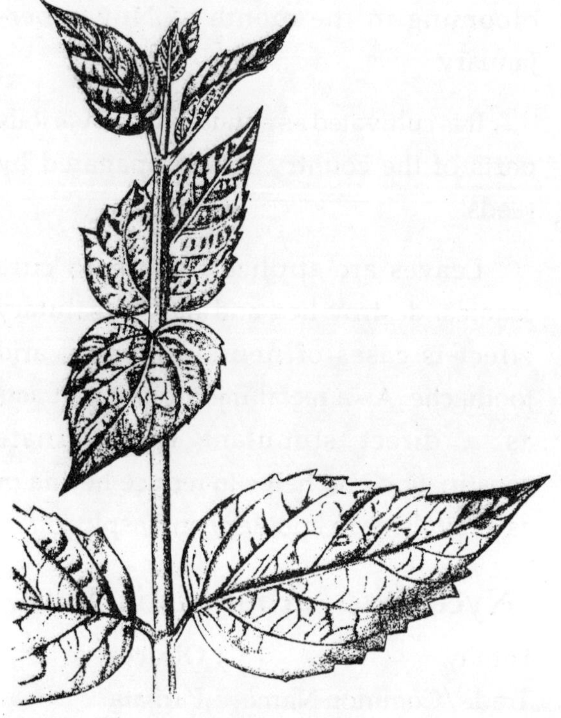
Nyctanthes arbor-tristis

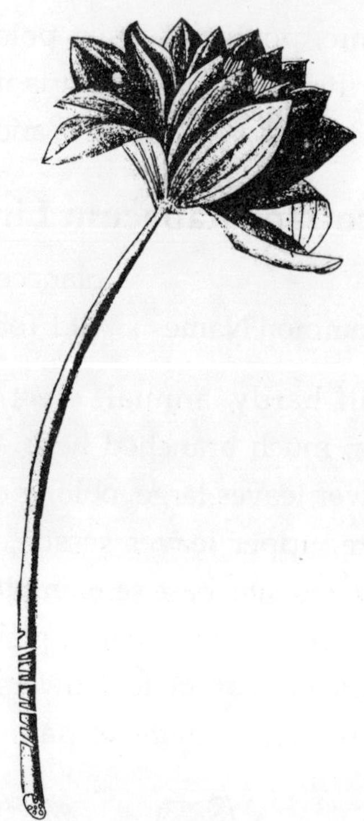
Nymphaea alba

benzoic acid. The leaves are antibilious, laxative, draphoretic, diuretic, cholagogue, mild bitter tonic and expectorant and are useful in fevers and rheumatism.

Nymphaea alba Linn.

Family : Nymphaeaceae
Trade/Common Name : Common water lily

A hardy aquatic, tuberous rooted, perennial herb. The rhizomes are stout, creeping in nature and at the tip it is bent up. Leaves are large, stipuled and float on surface which is nearly circular, entire and leathery. Flowers on long stalks and white in colour.

It is found throughout the India. It is propagated by divisions of clumps. Tubers or by cutting from rhizomatous stem.

It is administered against burning and insomnia and also improne the voice. It is used as emollients and demulcents. Locally, its pulp if used externally as a rubefacient.

Nymphaea stellata Willd.

Family : Nymphaeaceae
Trade/Common Name : Indian Red Water Lily

A perennial herb. Leaves are floating, smooth, diverging lobes. Flowers yellowish, floating carried on long stalks and blooming in the month of July to November.

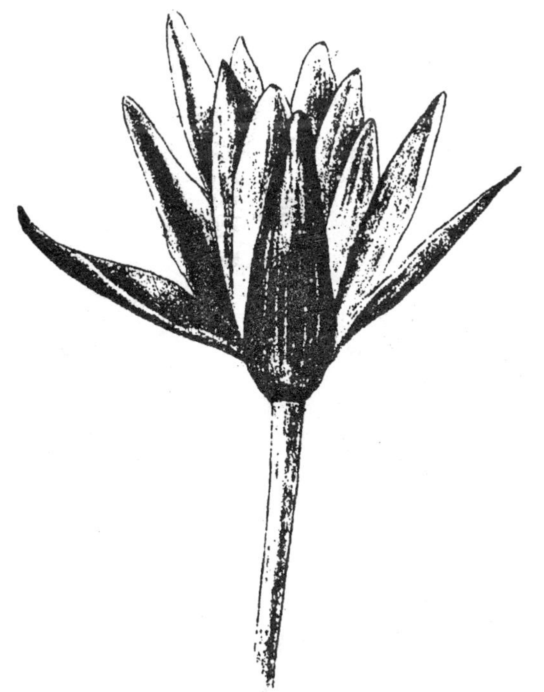

Nymphaea stellata

It is found shallow water in lakes, ponds and channels in Kashmir. It is propagated by cutting from rhizomatous stem.

Rootstalk given in dyspepsia, diarrhoea and piles. Decoction of flowers prescribed in palpitation of heart. The infusion of rhizomes and stem is used for blennorrhagia and diseases of urinary tract.

Ocimum basilicum Linn.

Family : Lamiaceae
Trade/Common Name : Sweet basil

An annual herb. Leaves numerous dot-like glands which secrete strongly scented volatile oils. Leaves are simple, petiolate, ovate or subovate, small, white flowers appear in clusters. Flowers are two lipped in raceme fashion.

Ocimum basilicum

It is native to India. It is cultivated in various parts of India. It is propagated by seeds.

Leaves and tender shoots yield a volite oil (0.4%). It contains methyl chavicol, methyl cinnamate, eugenol, linalool, camphor and geraniol. The herb possesses stimulating properties. An infusion of leaves is used to treat gonorrhoea, burning urine and nephritic affections. It is useful in the treatment of worms, stomach complains, fever, cough and gout. Leaf juice is employed as a nasal douche and also to treat skin diseases. The essential oil is cephalic and sedative properties. It is used in tonic neurosis.

Ocimum canum Sims.

Family : Lamiaceae
Trade/Common Name : Hoary Basil

A herbaceous plant. Bracts are petiolate. Flowers are more conspicuous. Seeds are black, ellipsoid and mucilaginous when wetted.

Ocimum canum

It is found in Jammu & Kashmir, Punjab, H.P. U.P., Uttaranchal and Delhi. It is propagated by seeds.

The herb yields an essential oil. Its constitutes are linalool and camphor. Seeds are tonic and diuretic.

Ocimum gratissimum Linn.

Family : Lamiaceae
Trade/Common Name : Shrubby Basil

An aromatic, much branched, under shrub or shrub, 50-120 cm high, young branches densely hairy. Leaves elliptic-lanceolate, cunsate at base, crenate-serrate, pubescent and gland-dotted on both surfaces. Flowers greenish-yellow, small, clustered in short, simple or branched, pubescent racemes and blooming in the month of September-October.

Ocimum gratissimum

It is found in plain & Central India. It is propagated by seeds.

Seeds and Leaves yield an essential oil containing eugenol (70-85%). It is used widely in perfumes, cosmetics, pharmaceuticals and confectionary industries. It is also used in bakery and ice cream industry.

Ocimum kilmand-scharicum Guerke.

Family : Lamiaceae
Trade/Common Name : Camphor Basil

A perennial undershrub. Bracts are petiolate. Flowers are more conspicuous. Seeds are black, ellipsoid and become mucilaginous when wetted.

It is introduced from Kenya in India. It is commercial cultivated in West Bengal, Jammu & Kashmir, Dehradun and Lucknow. It is propagated by seeds.

The herb yields an essential oil and a source of camphor. It is used in various preparation of ayurvedic medicine used for pains and sprains.

Ocimum kilmand

Ocimum sanctum Linn.

Family : Lamiaceae
Trade/Common Name : Holy basil

A much-branched, aromatic and erect herb, hairy throughout. Stem woody, perennial, giving out numerous branches. Leaves opposite, entire or serrate, toothed and dotted with minute glands. Flowers are purplish or crimson, small, in terminal, closely whorls racemes and blooming in the month of June-July.

It is found throughout India, ascending upto 1800 m in the Himalayas and in the

Ocimum sanctum

Andaman and Nicobar Islands. It is propagated by seeds.

Leaves and seeds yield an essential oil containing eugenol, carvacrol, methyl eugenol and caryophyllene. The oil possess antibacterial and insecticidal properties. Seeds are mucilaginous, demulcent, useful in complaints of the urinary systems. The juice or infusion of leaves possesses diaphoretic, antiperiodic, stimulating and expectorant properties. It is used in catarrh and bronchitis; applied to skin in ringworm and other cutaneous diseases and dropped into the ear to relieve earache.

Oenothera lamarckiana Linn.

Family : Onagraceae
Trade/Common Name : Primrosa

A biennial herb with submerged stems. Leaves simple, opposite, entire or toothed margin, stipules absent or caducons. Flowers solitary, auxillary or arranged in racemes.

It is found in tropical to sub-tropical Himayalan regions. It is propagated by seeds.

Seeds yields an oil (20-25%) containing gamma linolenic acid (GLA, 8-10%). The GLA rich oil is marketed as a health food which is used to treat various medical ailments and has therapeutic applications such as in cariovascular disorders, metabolic disorders, dermatological conditions, diabetes, breast problems, premenstrual syndrome, atopic eczema, rheumatic arthritis and alcoholism.

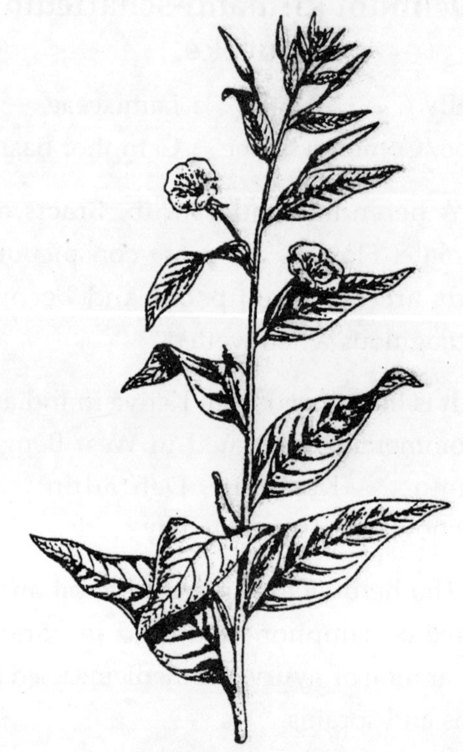

Oenothera lamarckiana

Onosma bracteatum Wall.

Family : Boraginaceae
Trade/Common Name : Ratanjot

A large, herbaceous perennial with black, woody rootstock, 2.5-5.0 cm in diameter, terminating in a knotty head from which arise several stems, erect or ascending. Root is purplish red from inside and the stem is simple and rarely branched, thickly studded with calcareous tubercles and armed with bristles. The leaves are entire, thick, petiolate and ovate-acuminate. Flowers deep blue, later turn purplish in

Mecanopsis aculeata

Mentha arvensis

Mentha longifolia

Mentha piperita

Mesua ferrea

Michella champaca

Mimusops elengi

Momordica charantia

Moringa oleifera

Mucuna pruriens

Musa paradisiaca

Nelumbo nucifera

Nerium indicum

Nyctanthes arbor-tristis

Nymphaea stellata

Ocimum basilicum

Ocimum canum

Ocimum kilmandscharicum

Ocimum sanctum

Oxalis corniculata

Panax pseudo-ginseng

Papaver somniferum

Pelargonium graveolens

Phyllanthus urinaria

Phytolacca acinosa

Picrorhiza kurroa

Piper betle

Pistacia integerrima

Plantago ovata

Piper longum

Plumbago zeylanica

Onosma bracteatum

heads, which are thickly covered by white, stiff, bristles and blooming in the month of August-September.

It is found in the Western Himalayas from Kashmir to Uttaranchal between 3600-4500 m elevation. It is propagated by seeds or splits.

Root yields a red dye containing mucilage and salt of Sodium, Potassium, Calcium, Iron and Magnesium. It is extensively used for colouring purposes. Bruised roots of Ratanjot are applied externally to cutaneous eruptions. The flowers of the plant are used as stimulant and cardiac tonic. Plant is used in rheumatism, syphilis and leprosy; useful for relieving of excessive thirst and restlessness in febrile excitement, in relieving heart palpitation, irritation of the bladder and stomach.

Onosma echioides (Gaertn.) Linn

Family : Boraginaceae
Trade/Common Name : Hairy Onosma

An erect, much branched hairy herb. Flowers are purple in colour.

Onosma echioides

It is native to Europe and Asia. It occurs rarely in Tehri Garhwal and in Alaknanda Valley, Dholi Ghatie, Kedar Nath and Madmaheshwar of Garhwal Hill between 3000-4000 m elevation. It is propagated by seeds or cuttings of root.

It is used in the Tibetan system of medicine. Locally, roots are applied on wound and cuts. It is also used as dye.

Operculina turpethum (L.) Silva Manoso

Syn. *Merremia turpethum* (L.) Shah & Bhatt.

Family : Convolvulaceae
Trade/Common Name : Turpeth

A large, climbing herb having milky juice with long, branched and fleshy roots. Leaves are ovate, cordate, upto 10 cm long and 7 cm broad. Flowers funnel shaped, appear few flowers in branches, white in colour.

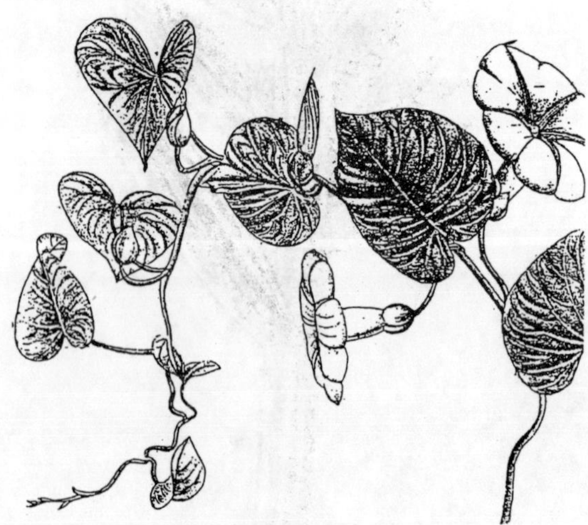

Operculina turpethum

It is native to tropical climate. It is widely distributed in India upto 1000 m altitude. It is propagated by seeds.

The dried roots with intact back has medicinal properties. It contains turpethin. Powered roots are used as purgative.

Ophioglossum reticulatum Linn.

Family : Ophioglossaceae
Trade/Common Name : Bourbon

A hardy deciduous fern. Fronds are barran, egg shaped, pale green, fertile ones contracted spike like.

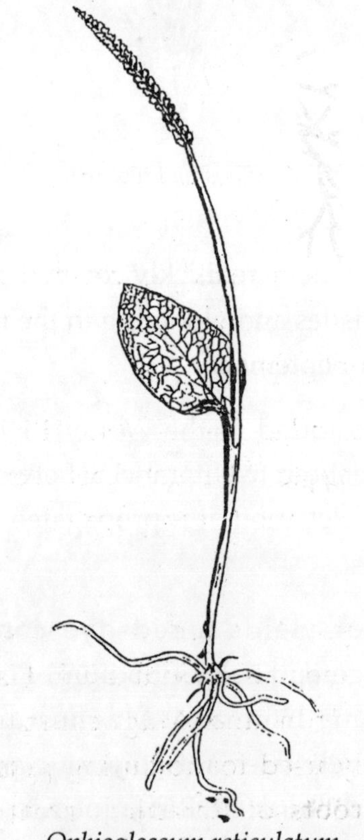

Ophioglossum reticulatum

It is native to Tropical Africa. It is uncommonly occurring in Mt. Abu of Rajasthan. It is propagated by spores.

The paste of the plant is applied on burns as cooling agent. The extract of leaf is also used in the preparation of tonic.

Origanum vulgare Linn.

Family : Lamiaceae
Trade/Common Name : Wild Marjornum

An aromatic, branched, perennial herb, 30-90 cm high. Leaves are broadly ovate and the flowers are terminal corymbose cymes with pale white colour or purplish or purplish tinge and blooming in the month of June to continue upto October.

Origanum vulgare

The herb is found in the temperate Himalayas from Kashmir to Sikkim between 1500-3000 m elevation. It is propagated by seeds, cutting, layers and root division.

Roots yield a volatile oil containing 50% thymol and also yield tannin and a bitter principle. The leaves and tops, cut prior to blooming are used to flavour foods. Oils possesses carminative, stomachic, diuretic, diaphoretic and emmenagogue properties. It is also given as a stimulant and tonic in diarrhroea, colic, and is applied in chronic rheumatism; toothache and earache.

Oroxylum indicum (L.) Vent.

Family : Bignoniaceae
Trade/Common Name : Arlu, Kinnauri phool

A small to middle sized, deciduous tree. Bark light grayish brown in colour. Leaves large, opposite, 3-pinnate near base, 2-pinnate in middle and simple pinnate above. Flowers large, purplish, fleshy in lax, terminal racemes as having 60-90 cm long peduncles and blooming in the month of June-July.

The tree is found in scrub forest and roadsides throughout the greater part of India, upto an altitude of 1200 m, mainly in

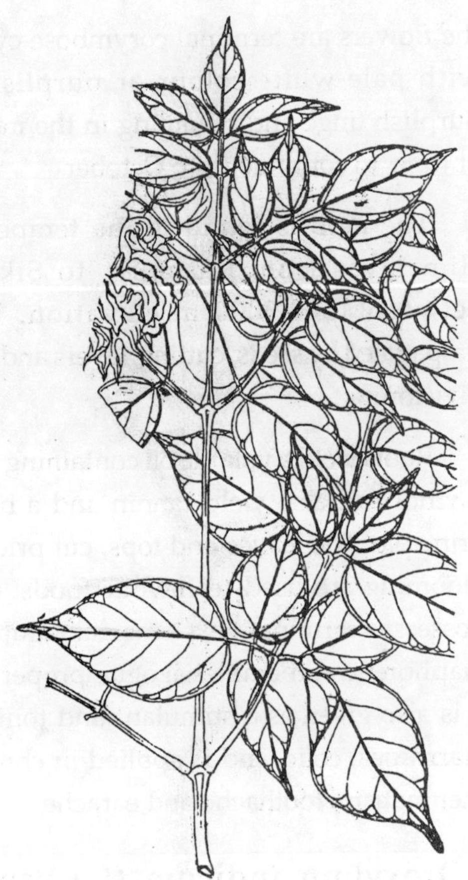

Oroxylum indicum

the ravines and moist places in the forests. It is propagated by seeds.

Leaves contain an anthraquinone derivative aloe-emolin, chrysin, baicalein, luteolin, 5,7-dihydroxy 6-methoxyllavone, etc. Bark is used as a tonic and astringent useful in diarrhoea, diaphoretic, rheumatism and dysentery. Tender fruits are refreshing and stomachic and the seeds are purgative. Leaves are used externally for enlarged spleen, headache and ulcers.

Oxalis acetosella Linn.

Family : Oxalidaceae

Trade/Common Name : Red Wood Sorrel

A stemless, perennial herb with rhizomatic creeping roots. Flowers are pinkish or whitish in colour.

It is found in th edump shady areas of Garhwal region of Uttaranchal upto 2500 m elevation. It is propagated by seeds, division of roots or offsets.

The leaves are eaten as salad. It is used to treat fever.

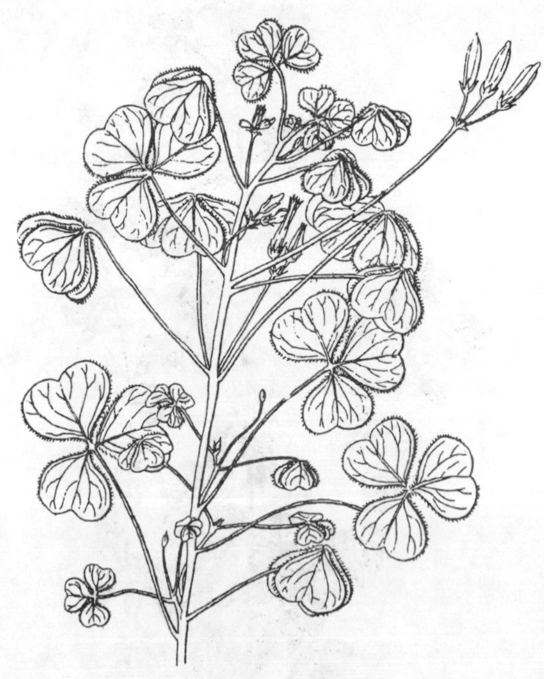

Oxalis acetosella

Oxalis corniculata Linn.

Family : Oxalidaceae
Trade/Common Name : Clover Sorrel

A delicate perennial hairy herb, procumbent, hairy with alternate, stalked, 3 lobed leaves. Flowers are pale pink or yellow, small in size and appear on long stalked infloresence.

It is found in different parts of H.P. upto 2800 m elevation, in shady places. It is also found in the damp and moist places upto 1800 m in Garhwal region of Uttaranchal. It is propagated by seeds, division of roots or offsets.

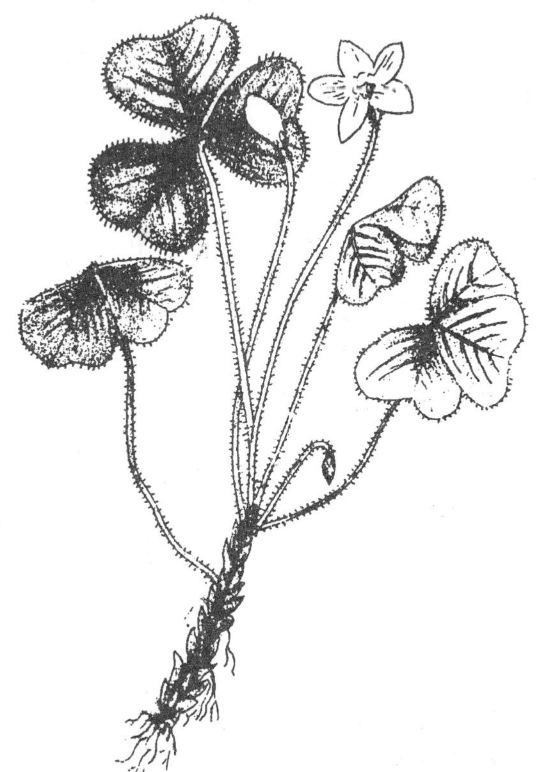

Oxalis corniculata

The herb is used as a cure for scurvy. It is good appetizer. Leaf juice is used locally in removing of the warts and also known as useful in cataract of the eyes. Paste of top shoots along with a few fruit of black pepper is applied to boil, absceses, wound and weeping eczima. Leaves and roots are used to treat dysentery and diarrhoea.

Paeonia emodi Wall. ex Royle.

Family : Paeoniaceae
Trade/Common Name : Himalayan Paeony

A glabrous, perennial herb. Stem 30-60 cm, leafy, erect. Leaves alternate, 15-30 cm long; leaflets 3; usually 3-parted, segments lanceolate, pointed, entire. Flowers few, slowly, 7.5-10 cm across, long-stalked, usually solitary in the axils of the upper leaves and blooming in the month of May-June.

It is found in North-West Himalayas from Kashmir to Uttaranchal between 1600-3000 m elevation. It is propagated by rhizome cuttings.

Rhizome contains starch (9.5%), sucrose (5.4%), malic acid (0.47%), oxalic acid (0.36%), tartaric acid (0.34%) and root oil possesses α-amyrin, butyropermol, cycloartenol, lupeol, linoleic acid, etc. Tuberous roots are used for intestinal diseases, colic, bilious obstructions, dropsy, epilepsy, convolutions and hysteria; given to children as blood purifier. Seeds are

Paeonia emodi

emetic. Plant is CNS depressant, hypothadermic, diuretic, anti-inflammatory.

Panax pseudo-ginseng Wall.

Family : Araliaceae
Trade/Common Name : Indian Ginseng

An erect perennial herb upto 60 cm tall. Leaves are compound, palmate. Petiole is longer. Leaflet are usually 5, sometimes 3, upper 3 layers upto 15 cm long and lower 2 upto 4 cm long, elliptic or ovate. Infloresence simple, umbelliform. Flowers greenish-yellow. Fruit berry-like. Root are thick, fleshy, short, cylindrical or fusiform.

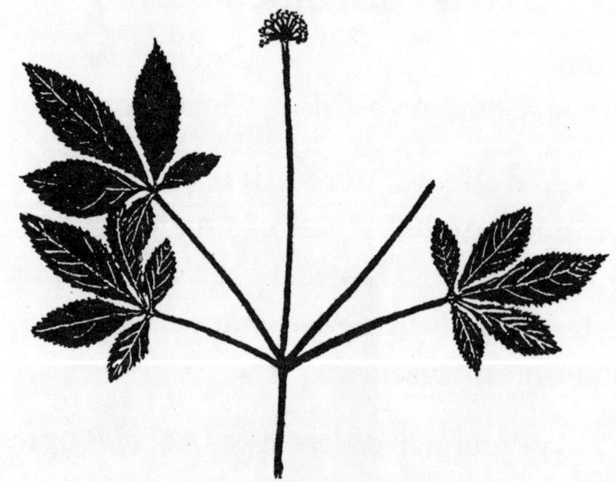

Panax pseudo-ginseng

It is found endemic in Skkim Himalaya between 1800-3600 m elevation. It is propagated byroot cuttings or grafting.

The root constitute the drug which contains panaxosides, steroids, vitamins and trace elements. It is tonic, stimulant and aphrodisiac in nature. It is increased the natural resistance and recuperative power of the body. It is used for the treatments of amnesia, headache, convulsion, dysentery and cancer. It is also prescribed for neurasthenia, dyspepsia, palpitation and asthma.

Pandanus fascicularis Lam.

Syn. *P. odoratissimus* L.f.

Family : Pandanaceae
Trade/Common Name : Kewda

A hardy, much branched, bushy-shrub or small tree, 2-6 m high; stem supported by basal stout, pro roots. Leaves linear-ensiform, caudate-acuminate, drooping,

Pandanus fascicularis

glossy green, margins with uncarved spines. Flowers pale-yellow, sweet scented, in axillary clustered, spathe creamish-yellow, fragrant. Male flowers of numerous anther, umbellate on the connate filament and blooming in the month of February-April.

It is found in waste-lands and marshy lands of Central India. It is propagated by seeds or cuttings.

Male spadices yield an essential oil known as 'Kewda oil'. It contains methyl ether or phenyl alcohal (66.68%). The male flowers are highly fragrant and are used for the preparation of 'kewda attar' which is among the ancient perfumery articles of India. It is used in almost all types of perfumes and is used for scenting oils, tobacco and agarbatti. The juice of the infloresence is useful in rheumatic arthritis in animals.

Papaver somniferum Linn.

Family : Papaveraceae
Trade/Common Name : Opium Poppy

An erect, annual herb upto 1.2 m tall. Leaves large, serrated. Flower are red, pink white, purple in clour and blooming in the month of April to July.

Papaver somniferum

This herb is grown in India commercially as annual crop during winter in restricted areas to a contiguous tracts in few districts of Madhya Pradesh, Chhatisgarh, Rajasthan, and Uttar Pradesh. It is propagated by seeds.

The latex of immature fruit contain several alkaloids like morphine, codeine, noscapine, narceine and papaverine. It is used as pain killer, cough, depressant, sedative, antitussive and smooth muscle relaxant. Seeds are important source of oil, valued for culinary purposes.

Peganum harmala Linn.

Family : Zygophyllaceae
Trade/Common Name : Syrian rue

A perennial, much branched bushy herb. Leaves somewhat fleshy, pinnately cut into linear acute lobes. Flower solitary in terminal axils of small branches, greenish-white in colour.

It is found in wastelands, orchards, rocky substrate, dry areas in North-West India. Prominent in Indus velleys from 100-5000 m altitude. It is propagated by seeds or cuttings.

Seeds contain harmaline, harmine and The herb possesses aphrodiasic, abortifacient, amoebicidal, anodyne, diuretic properties. Seeds are narcotic and given in fever and also used as a remedy for tapeworm in man. It is used in the treatment of jaundice, fever, painful menstruation, colic pains, gallstones, asthma, hysteria and rheumatism.

Peganum harmala

Pelargonium graveolens (L.) Hervitt.

Family : Geraniaceae
Trade/Common Name : Geranium

A hardy slender herb. Leaves palmately lobed and densely pubescent. Flowers small and arrange in a panicle, borne on long peduncles with light or deep pink in colour.

It is native of South Africa and introduced in India at Nilgiris (1954). It is grown in South Indian hills and at lower altitudes of Kashmir and Jammu and Uttaranchal. It is propagated by seeds and shoot cuttings.

Leaves yield an essential oil known as geranium oil. It possess a strong rose-like odour. The chief constitutes of oil are geraniol (68-75%) and citronellol (23-40%). The oil is ued in many perfumery and cosmetic industries due to its agreeable odour. Tannin can also be obtained as a bye-product from its stem and leaves after extraction of oil.

Pergularia daemia (Forsk.) Chiov.

Family : Asclepiadaceae
Trade/Common Name : Pergularia

A hairy climber with milky juice. Leaves are ovate or round, upto 6 cm long, lower surface is hairy. Small flowers appear in short clusters, yellish green in colour.

It is native to Madagascar. It occurs in almost all parts of India upto 1500 m altitude. It is propagated seeds.

The plant is emetic, purgative, expectorant and used as uterine tonic. Leaf juice is used to treat infantile diarhhoea, ashma and catarrhal affection, rheumatism and amenorrhoea. It is also useful in curing gynecological conditions like excessive bleeding. Squeezed fresh leaves are applied as poultice in carbuncle.

Perilepta auriculata Nees.

Family : Acanthaceae
Trade/Common Name : Kapurminjri

A shrub with pale blue tubular flowers. The fruting brownish hairy spike bears oil droplets which emit camphorous odour.

It is native to Indo-China. It is found in kangra and Hamirpur districts of Himachal Pradesh. It is propagated by seeds.

The oil droplets from the fruting spike are mixed with kajal (Collyrium) and is applied to the patient suffering from cataract and trachoma.

Permelia perlata (Huds.) Ach.

Family : Parmeliaceae
Trade/Common Name : Stone Flower

A thallus of dirty white or gray colour, 5-10 cm long and made up of numerous fibres spreading from common stalk upon the tree or rock, often forming a rosette like flat mass. Fibres are flat, generally dichotomously branched, marked here and there irregular depressions.

It is commonly found growing on rocks and on the trunks and branches of oak trees between 1600-3200 m elevation.

The herb contains lecanoric acid and atronorin, a yellow crystalline matter called chrysophanic acid. It is demulcent, tonic, febrifuge and diuretic. It is given in amenorrhoea, chlorosis and leucorrhoea. As a soporific and sedative, it is given to stop nocturnal emissions. Power is used as a snuff to remove headache. It is useful in stomach disorders, dyspepsia, vomiting, pain in the liver or womb. Lecanoric acid is an excellent material for the manufacture of litmus.

Pholidota articulata Linn.

Family : Orchidaceae
Trade/Common Name : Rattle Snake Orchid

A epiphytic, annual plant with pseudobulb are jointed, 2 leaves, cylindric upto 10 cm long. Leaves are thickly membranous, many nerved, elliptic, acute, short-stalked to 10 cm long and 4 cm broad. Flowers musk scented, yellowish-white in colour and appearing on long drooping.

It occurs in Uttaranchal, Sikkim, Manipur and Khasia hills between 700 to 1300 m altitude. It is propagated by seeds or bulbs.

The whole plant is used as tonic.

Phyllanthus fraternus Webster

Syn. *P. niruri* Auct. (non L.)

Family : Euphorbiaceae
Trade/Common Name : Bhuiamla

An erect, much branched, annual herb upto 45 cm in height. Leaves small, simple, appears to be compound, elliptic-oblong, distichous. Flowers very small, creamy-white, axillary, hanging down below the branchlets, male 1-3; female solitary and blooming in the month of September – October.

Phyllanthus fraternus

It occurs sporadic in wastelands in suntropical regions upto 1000 m elevation. It is propagated by seeds.

The herb possess astringent, stomachic, cholagogue laxative, deobtruent and bitter tonic properties. It is used as a remedy for dropsical disorders, gonorrhoea and other genitourinary diseases, jaundice, constipation, stomachache, dyspepsia and dysentery. Locally, its juice is used in case of liver disorder.

Phyllanthus urinaria Linn.

Family : Euphorbiaceae
Trade/Common Name : Lalbhuin

An annual, erect or decumbent-ascending herb, to 70 cm high; branches tinged with red; branchlets flattened or slightly winged. Leaves sessile, distichous, linear-oblong or oblong, base cuneate, rounded, slightly oblique, apex obtuse or

Phyllanthus urinaria

Physochlaina praealta

rounded, mucronate; dark green above, pale-green below; stipules peltate, ovate-triangular, acuminate. Flowers minuta, female solitary, male 1-3 in the axil of leaves and blooming almost throughout the year.

It occurs in the hill slopes of Uttaranchal and also in other parts of India. It is propagated by seeds or cuttings.

It is used to treat liver diseases and jaundice. The whole plant is diuretic.

Physochlaina praealta (G. Don) Miers.

Family : Solanaceae

Trade/Common Name : Laltang

An erect, glabrous, perennial herb, 150-160 cm high. Leaves are irregular, ovate-oblong, wavy; flowers greenish, campanulate, in terminal corymbose cymes and blooming in the month of June to continue upto August.

The herb is found in frequently as a roadsides weed often forming patches running and covered large area in Kashmir, Himachal Pradesh and Western Tibet at altitudes of 3300-4600 m. It is propagated by seeds.

Leaves contain an alkaloid-tropane (1.02%), and also having hyoscyamine, atropine, hyoscine and sucrose. It is narcotic and possesses mydriatic properties, causing dilation of the pupil of the eye, like belladonna. The seeds are used by local people as vermifuge to expel round worms and as emetic in bilious attack.

Physostigma venenosum Balf.

Family : Fabaceae
Trade/Common Name : Ordeal Bean of Calabar

A woody climber with branches. Leaves are trifoliate. Leaflets rounded at base and pointed at the apex. Infloresence axillary appears in pendent clusters. Flowers are kidney shaped, purple in colour.

It is native to west Tropical Africa. It occurs in coastal areas of Southern India and Nilgiris hills. It is propagated by seeds.

Seeds contain starch (48%), albuminoids (23%) and several alkaloids like physostigmine, eserdine, calabarine and physovenine. It is used to treat skin diseases, rheumatism and the liquid extract of seeds is used as a sedative. Physostigmine salicylate is admintered orally, subcutaneously or intramuscularly as an antispasmodic in rheumatoid arthritis, bursitis and fibrostitis. Physostigmine is also used in ophthalmology, it also acts as a strong sedative on the spinal cord. Eseridine is a purgative.

Phytolacca acinosa Roxb.

Family : Phytolaccaceae
Trade/Common Name : Sweet Belladona

A succulent, robust, glabrous, erect herb upto 1.5 m tall. Leaves alternate, broadly lanceolate, 15-25 cm, entire, long pointed, narrowed into a short stalk. Flowers pale green, aggregated in leaf opposed, robust, cylindrical racemes.

Phytolacca acinosa

It is found in forest shades of temperate Himalaya from Kashmir to Bhutan and in Assam in between 2300-2800 m elevation. It is propagated by seeds and root splits.

Seed kernels yield a fatty oil. The tender leaves and twigs are cooked as vegetables. The herb is believed to have narcotic effect, which is destroyed on boiling. The fruit is occasionally eaten and used for flavouring purposes.

Picrorhiza kurroa Royle ex Benth.

Syn. ***Picrorhiza scrophulariflora*** **Pennell.**

Family : Scrophulariaceae
Trade/Common Name : Kutki

A trailing herb with jointed and zig-zag

Picrorhiza kurroa

underground stems (rhizomes), giving off shoots at joints and also rooting at the nodes. Leaves are spathulate and denate. Flowers are bluish white, showy and are arranged in dense, terminal and spicate racemes and blooming in the month of June to continue upto August.

It is found in rocky slopes, grassland in the alpine Himalayas from Kashmir to Sikkim, at the elevation of 3000-4000 m. It is propagated by rhizome, stolons and offsets.

Rhizome contains a brown resinous glucoside, picrorhizin and picrorhizetin. It also possesses kutkin, apocynin, α-sitosterol. It is considered to be anthalmintic, appetizer, blood purifier, blood pressure reducer, cardiac, expectorant, ferbifuge and useful in asthma, cold and cough, bile trouble, jaundice, leprosy, constipation and stomach troubles.

Pimenta officinalis Linn.

Syn. *P. dioica* (L.) Merr.: *Eugenia pimenta* DC.

Family : Myrtaceae
Trade/Common Name : Allspice

An evergreen, aromatic tree upto 12 m height. Leaves are aromatic. Flowers are white in colour. Fruits are berries, globular, raddish brown in colour.

It is native to West India. It is cultivated in Jamaica and Guatemala. It is grown in the gardens of Bangalore, Bihar, Orissa, Kerala and West Bengal. It is propagated by seeds or cuttings.

Fruits yield an essential oil which contains eugenol, phellandrene,

Pimenta officinalis

caryophyllene, methyleugenol and cineole. It is carminative in nature. The berries are used as condiments for flavouring ingredient. It possesses stimulant, tonic, purgative, digestive and carminative properties. The podwered fruit is used to treat flatulence, diarrhoea and dyspepsia. It is also used againt neuralgia and rheumatism.

Pimpinella anisum Linn.

Family : Apiaceae
Trade/Common Name : Vilaiti Saunf

A hardy annual, pubescent, aromatic herb of 60 cm tall. The basal leaves are long and dissected with long petioles. Leaves on stem are pinnately decompound. Flowers appear in umbels, small, white in colour and borne centripetally.Fruit is small, ovoid, laterally compressed, grayish green in colour. After drying, the fruits become grayish brown.

It is native to Greece to Egypt. It is commercially cultivated as a spice crop in various parts of India. It is propagated by seeds.

Seeds contain essential oil (2-3,5%) which constitute anethole (90%) and also nethylchavicol, p-nethoxyphenyl-acetone and acetic aldehyde. It is used as cadiments. It has carminative, diuretic and diaphoretic properties. It is used externally in the form of fomentations, lotions and poutices for lacteal ecchymoses and obstructions. Fruits are regarded as a stimulant, carminative, stomachic, diuretic, expectorant and emmenagogue. It is used in the debility of digestive organs, gastralgia, flatulence, windly and spasmodic colics, dyspepsia, giddiness, dizziness and nervous headache.

Pimpinella anisum

Pinus roxburghii Sarg.

Family : Pinaceae
Trade/Common Name : Chir Pine

A moderately tall tree with a spreading crown; branches more or less whorled, bark dark gray, often reddish, deeply fissured and rough. Leaves are long, 20-30 cm acicular, triquetrous, light green in clusters of male flowers arranged in the form of cones and blooming in the month of April to continue upto June.

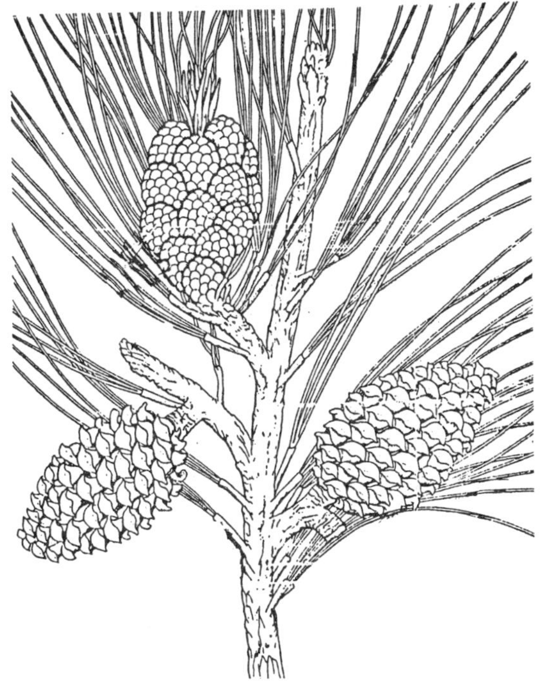

Pinus roxburghii

It occurs vast stretches of forests in Uttaranchal and Himachal Pradesh between 450-2200 m elevation. It is propagated by seeds.

Oleo-resin yields turpentine oil and rosin. It is used in the pharmaceutical preparations, perfumery industry, synthetic pine oils, insecticides and denaturants. It is feebly antiseptic and as expectorant is useful in chronic bronchitis. As nema, the oil is useful in obstinate, constipation, tympanitis and seatworm infestations.

Pinus sylvestris Linn.

Family : Pinaceae

Trade/Common Name : Wild Pine

A tree with spreading branched attain up to 30 m height. Leaves are dense, of a glabrous hue, up to 5 cm long. Leaf sheaths are small, persistent, nearly black. Grey green needles are in pairs. Cones are solitary or two or three together, about 5 cm long, tapering towards the apex. Seeds are winged. Bark is reddish.

It is native to Europe. It is found in hilly parts of India. It is propagated by seeds or cuttings.

The young branches and the needle yield 0.25 to 0.35% volatile oil with pleasant balsamic odour. It contains pinene, limonene, careen, camphene and sesquiterpene alcohols. The oil is used as an adjunct in flavouring deodorants,

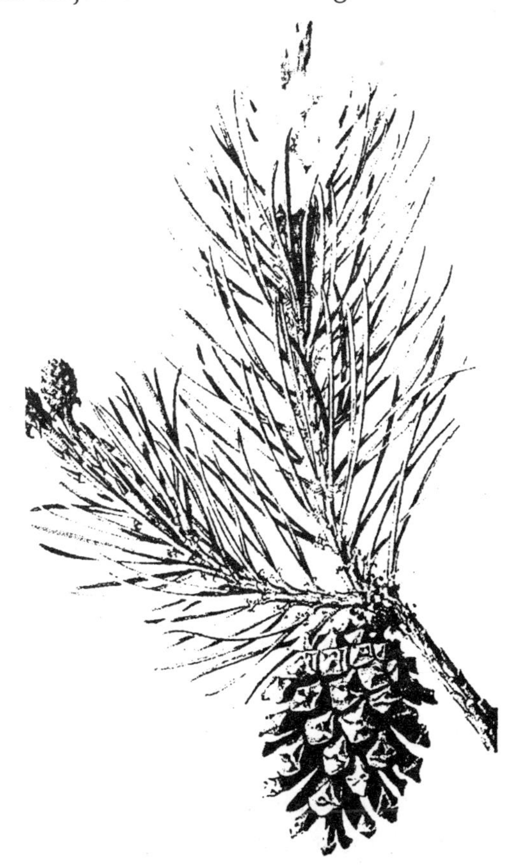

Pinus sylvestris

inhalants, room sprays, soaps and bath preparations. Fresh cones are boiled in whey and beer brewed with the tender hops of the branches and is said to be good in inveterate scurries.

Pinus wallichina A.B.Jackson.

Family : Pinaceae
Trade/Common Name : Blue Pine

A tall, evergreen tree with blue-green foliage. Leaves (needles) 12-20 cm long in clusters of 5, drooping, except when quite young. Male cone ovoid, short. Female cones on maturity are cylindric, 15-25 cm long, erect when young, pedulous afterwards and blooming in the month of May to continue upto August.

It is found in the form of good forests either pure or mixed with deodar, fir, spruce and broad leaved trees of *Acer, Aesculus, Prunus* and *Quercus* between 2000-3500 m in Uttaranchal and Himachal Pradesh. It is propagated by seeds.

Oleoresin yield turpentine oil, about half of chir pine, but superior quality and high pinene contents. Needle oil shows antibacterial properties. The resin from the young sapling is used as liniment.

Piper betle Linn.

Family : Piperaceae
Trade/Common Name : Betelvine

A perennial bisexual herb. Branches arises from the stem nodes. Nodes thick and

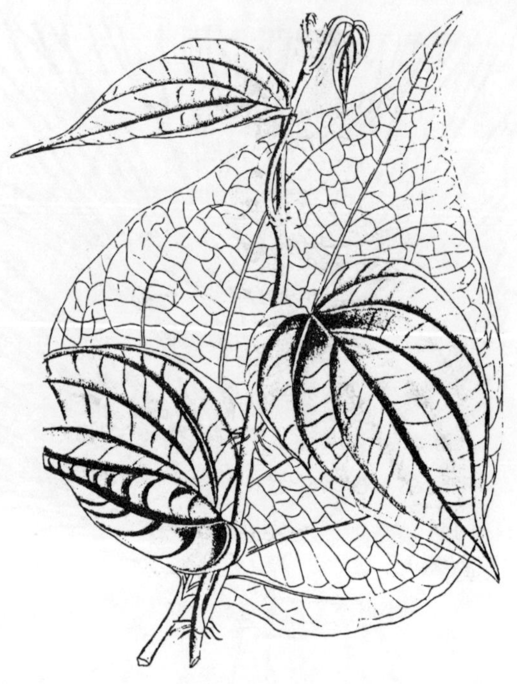

Piper betle

smooth. Flowers in sympodial spikes, the bracts closely oppressed to the axis.

It is native to Centeral and East Malaysia and India. It is cultivated as commercial crops in South Indian parts. It is propagated by cuttings.

The leaves yeild an essential oil with characteristic smell and their hot taste . It also contains vitamin B and C. It stimulates brains, lungs and heart.

Piper longum Linn.

Family : Piperaceae
Trade/Common Name : Long pepper

A creeping aromatic herb. Lower leaves are broadly ovate, deeply cordate with big lobes at base upto 10 cm long, upper leaves

Piper longum

dark green oblong-ovate, cordate. Flowers are in spikes, bracks or male spikes narrow and female spikes circular.

It grows in tropical zones and distributed in Central Himalaya, Khasi and Mikir hills, Western Ghats, lower hills of W. Bengal and Assam. It is cultivated in Assam and Tamil Nadu. It is propagated by seeds.

The fruits contains piperine and piplartine alkaloid; yields 0.7% essential oil with spicy odour. In medicine system, the root is used as a carminative, tonic to the liver, stomachic, emmenagogue, abortifacient and aphrodisiac. It is effective to treat insomnia, epilepsy, obstruction of bile duct and gall bladder, dysentery and leprosy. The fruits possess haematinic, diuretic, digestive, tonic properties and useful in inflammation of the liver, pains in the joints, lumbago, snakebite, scorpion sting and night blindness.

Piper nigrum Linn.

Family : Piperaceae
Trade/Common Name : Black pepper

A perennial climber herb. Leaves large, broad, ovate and round. Flowers appear in long spikes. Fruit, a berry gathered from ripe and dried, forms a black pepper corn.

It occurs in hot and moist parts of Southern India. It is propagated by root cuttings.

The Fruit contains an crystallizable alkaloid piperine with biting taste. It yeilds 1.0-2.5% volatile oil. It possess anti-blennorrliagic, stomachic, stimulant, aphrodisiac and carminative properties. It is useful in the treatment of dyspepsia, malaria, haemorrhoids, tremors and

Piper nigrum

delirium. Locally, it mixed with cinnamon is used for treating migraine. It is used in confectionary, beverages, canning, pickling and baking, seasoning of dishes etc.

Pistacia khinjuk Stew ex. Brandis

Syn. *P. integerrima* Stewart.

Family : Anacardiaceae
Trade/Common Name : Kakrasingi

A moderate sized deciduous tree with rough, gray bark. Leaves odd or even pinnate long and the leaflets are nearly opposite, lanceolate, shining, entire. Flowers are uni-sexual and deciduous and blooming in the month of March to continue upto May.

Pistacia khinjuk

It is found in the Himalayas from Indus to Uttaranchal at the altitude of 400-2350 m. It is propagated by seeds.

The galls are found on the leaves especially the petioles by the insects and are called as 'kakarsingi'. The galls contain tannins, a resin and an essential oil (1.3%). The essential oil containing α-pinene, camphene di-limonene, cineole, a resinous substance. The galls are used for tanning and dyeing, besides being used in asthma, pthisis and other diseases of the respiratory tract and in dysentery. The oil, found in the galls, is used as a carminative.

Plantago lanceolata Linn.

Family : Plantaginaceae
Trade/Common Name : Wild Isabgol

A perennial, glabrous herb. Leaves all radical in a rosette, ovate-lanceolate, entire, base decurrent into short petioles, 3-5 nerved. Scapes few, 30-60 cm high, deeply ribbed; spikes compact, shortly cylindrical, pale-white in colour and blooming in the month of April to continue up to September.

It is found in grasslands, waysides, wastelands, orchards, as weed in cultivated field from Kashmir to Himachal Pradesh in western Himalaya. It is propagated by seeds.

Leaves and roots possesses astringent properties and used for coughs and

Plantago lanceolata

Piper ovata

pulmanory disorders. Leaves applied on wounds, inflamed surfaces and sores. Alcoholic extracts of young leaves exhibit antibacterial action. Seed used with sugar as a purgative.

Plantago ovata Forsk.

Family : Plantaginaceae
Trade/Common Name : Isabgol

A small, stemless herb covered with dense or soft hairy growth. Leaves narrow, long. Flower small, oval or cylindrical spikes. Seeds boat shaped, pinkish gray brown in colour. Each seeds is encased in a thin, white, translucent membrane called husk.

In India, it is cultivated in Gujrat and adjoining areas of Southern Rajasthan, Punjab and Uttar Pradesh. It is propagated by seeds.

The husk of seed yields a colloidal mucilage, consists of xylose, arabinose and galacturonic acid. The mucilage is over 30% of the whole seed. It serves as a safe bulk-laxative, promotes regular bowel movement. For chronic dysenteries of amoebic and basillary origin. It is used to cure inflammation of the mucous membrane of gastro-intestinal and genito-urinary tracts, dueodenal ulcer, gonorrhoea and piles. It is also used as cervical dilator for termination of pregnancy.

Pleurospermum brunonis (DC.) Benth.

Family : Apiaceae
Trade/Common Name : Nesar

A small tufted herb upto 20 m tall, highly scented. Leaves are finely dissected. Flowers are dark brown in compound umbels, closed by the creamy white bracts.

It is sporadically grows on rock bounders, rocky slopes and rock crevices in sub-alpine and alpine zones between 3500-4500 m elevation in Himachal Pradesh. It is propagated by seeds or cuttings.

The powder of flowering shoots is mixed in cow's fresh butter and massaged over the entire body to allay fevers. The dried herbs is kept in the boxes containing clothes as a preservative against attack of moths.

Plumbago zeylanica Linn.

Family : Plumbaginaceae
Trade/Common Name : Leadwort

A rambling, subscandent perennial herb or under shrub with woody, striate stem. Leaves ovate, glabrous and the flowers are white in colour in elongated spikes and blooming round the year except winter.

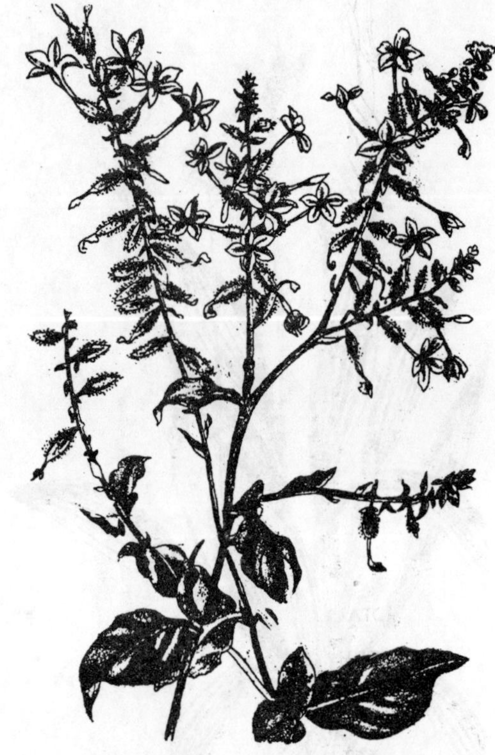

Plumbago zeylanica

It is widespread and is indigenous to South-East Asia. It is found wild in peninsular India and West Bengal and cultivated in gardens throughout India. It is propagated by seeds.

The root contains a number of naphthaquinone derivatives. Leaves and stem contain volatile oil. The root possesses obortifacient and vesicant properties. It is used as diuretic, caustic and expellent of phlegmatic tumors and is useful in rheumatism. It is used as an irritant of the skin, in the treatment of dyspepsia, piles, anasarca, diarrhroea and skin diseases.

Podophyllum hexandrum Royle

Syn. *P. emodi* Wall. ex Royle

Family : Podophyllaceae
Trade/Common Name : Indian Podophyllum

An erect, glabrous, succulent herb; 30-60 cm tall with creeping, perennial, knotted rhizomes bearing numerous roots. Leaves usually 2; long, stout, stalked, orbicular-reniform, palmate, peltate with lobed segments. Flower solitary, in the axil of the two leaves; creamy white, cup shaped and blooming in the month of April to continue upto May.

It is found in open slopes, forest clearings, moist humic places from Kashmir to Uttaranchal in the temperate Himalaya between 2600-4500 m. It is propagated by cutting.

Podophyllum hexandrum

It yields an resin (7-15%), podophyllin, podophyllotoxin, podophyllol, quercetin, kaempferol, astergalin, an essential oil. The roots are richer in resin than the rhizome. It lost considerable after long storage. The rhizome and roots are considered as hepatic, stimulant, cholagogue, purgative, alterative and bitter tonic. The resin or podophyllin is a drastic purgative and produces copious watery stools. It is an effective vermifuge, first stimulating, then paralyzing and finally killing ascaris. During recent years, podophyllin has aquired special importance for its possible use some forms of cancer.

Pogestemon patchouli Hook.

Family : Lamiaceae
Trade/Common Name : Patchouli

A hairless herbaceous plant with 1 m height. Twigs are quadrangular with opposite leaves. Leaves are of a pale green colour, heart-shaped, slightly lobed and with downy hairs and obtuse lobes and apex and also having glands which secrete the real oil.

It is grown in coastal areas of South India beside hill stations upto an elevation of 1500 m. It is propagated by seeds or shoot cuttings.

Whole plant (i.e. roots, stem, branches, stalks and leaves) contains essential oil (3-3.5%) but the oil contents of the leaves is by fat the highest. The oil contains Patchoulol (30-40%). The oil is used as a 'base' material in perfumery industry as it has strong fixative properties. It is also used as a flavour ingredient in many food products. It also possesses many medicinal properties.

Polianthus tuberosa Linn.

Family : Amaryllidaceae
Trade/Common Name : Tuberose

A tuberous root stock from which grows an erect flower stem 90 to 120 cm high. It produces mass of grass like foliage, arching in shape, pale green, long, narrow and very dense. Flowers are funnel shaped.

It is native to Central America and is widely cultivated in India. It is propagated by bulbs.

Flowers yield an essential oil which contains methyl benzoate, methyl anthrinilate, benzyl alcohol etc. It is considered to be hot, dry and diuretic. Dried powdered bulbs are used as a remedy for gonorrhoea. To remove small red pimples they are rubbed up with turmeric and butter.

Polianthus tuberosa

Polygonatum multiflorum (Linn). All

Family : Liliaceae
Trade/Common Name : David's Herp

A hardy, perennial herb upto 1 m tall.

It is native to Europe. It occurs commonly in Bhatanu of Uttar Kashi between 2000-3000 m. It is propagated by divisions of bulbs.

Polygonatum multiflorum

The paste of seeds and leaves is applied in itching and chopped skin.

Polygonatum verticillatum (Linn). All

Family : Liliaceae
Trade/Common Name : Meda

An erect or sunerect, perennial herb with creeping, gingerlike rootstock, 60-120 cm high. Leaves sessile, in whorls of 4-8, linear of linear-lanceolate in shape. Flowers greenish-white, on 2-3 flowered peduncles, forming axillary whorls and blooming in the month of May to continue upto August.

It is distributed frequently as undergroth in forest of temperate regions between 2000-3000 m elevation. It is propagated by root cuttings.

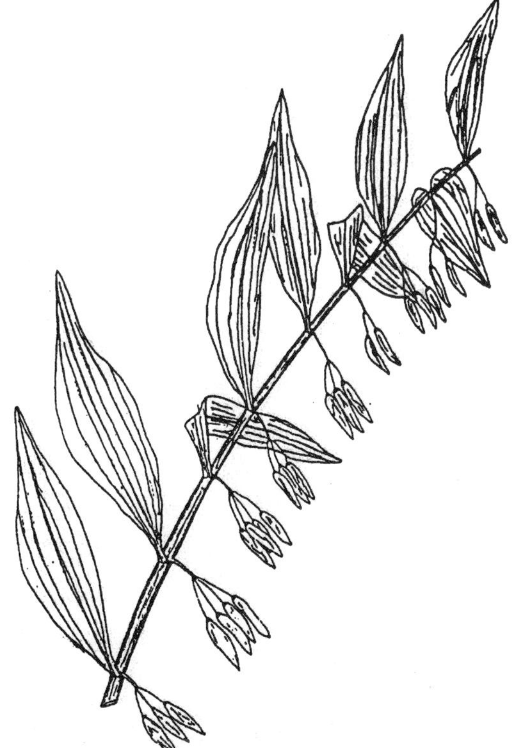

Polygonum verticillatum

The rhizomes are considered to be one of the eight ingredients of 'Astavarga' of Ayurveda. Is is used as tonic in various ayurvedic medicine.

Polygonum alpinum All.

Family : Polygonaceae
Trade/Common Name : Chita-hola

A perennial herb upto 1.5 m tall, jointed stems. Leaves oblanceolate with long stipules. Flowers white in much branched narrow pyramidal clusters and blooming in the month of June-July.

It is found shrubberies, alpine grassland, streamed sides and forest clearing from Kashmir to Himachal Pradesh in the western Himalaya. It is propagated by root cuttings and seeds.

Rootstock contains tannins (16-22%), soluble non-tannins (14-18%) and sugars (5%). It is used to be potent home remedy against all joint pains especially backache and control dysentery in calves and fawns.

Polygonum aviculare Linn.

Family : Polygonaceae
Trade/Common Name : Knotgrass

A highly variable annual herb, with prostrate stems and ascending upto 60 cm long having partly purple tinted colour. Roots spindle at top and branched at

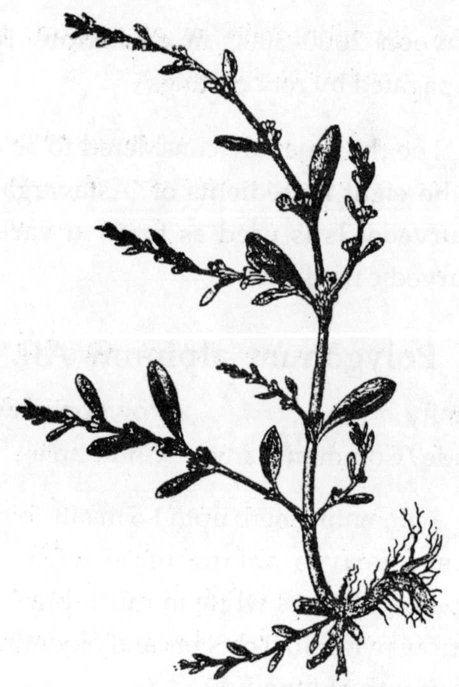

Polygonum aviculare

bottom, producing several, many branched, jointed stems that usually spread in every directions. Leaves elliptic or lanceolate, set at nodes and right upto the tip of branches. Flowers small, green, tipped with red or white and in axillary clusters and blooming in the month of June-July.

This herb is found sporadic in temperate and sub alpine areas from Kashmir to Uttaranchal at altitudes of 1800-3600 m. It is propagated by seeds.

The herb is a source of Vitamin E and contain tannins, mucilage, sugar, gallic, caffeic, sillicic, chlorogenic and p-coumaric and several other flavonoids. It is possess astringent, tonic, antipyretic, antiseptic, diuretic, haemostatic and vermifuge properties. The seeds are aromatic, and powerfully emetic and cathartic. It is used as a uretic for gall bladder and kidney problems and for rheumatic complaints.

Polygonum hydropiper Linn.

Family : Polygonaceae

Trade/Common Name : Water pepper

A perennial herb upto 50 cm tall with greenish-yellow leafy. Leaves oblong-lanceolate with characteristic pepper taste, stipules prominent concealing the stalk. Flowers are small, sessile in terminal, generally unbranched spikes and blooming in the month of June-July.

It is found damp places, ditches, wetlands, shallow water ponds from North-west India to East. It is propagated by seeds.

Polygonum hydropiper

Leaves yield an essential oil and also contains oxymethyl-anthraquinones, polygonic acids. The herb possesses bitter, stimulant, tonic, diuretic, carminative, anthelmintic, emmenagogue, haemostatic and lithotriptic properties. Locally, its liquid extract used as an oral contraceptive and infusion given in uterine disorders. Leaves chewed to relieve toothache.

Pongamia pinnata (Linn.) Merr.

Family : Fabaceae
Trade/Common Name : Indian Beech

A handsome tree with drooping branches; leaves ovate, acuminate shinning green leaves laden with rose tinted white flower, arise in axillary racemes and blooming in the month of April to continue upto May.

The plant is found throughout India, very common as a sea-coast tree in the Andaman and Nicobar Islands. It is propagated by seeds, root sucker and root-shoot cutting.

Seeds contain a bitter fatty oil (17-36%) and traces of an essential oil and three crystalline substances-karanjin, pongamol and glabrin. The oil is antiseptic, useful in skin diseases. Bark is astringent, carminative, cooling, demulent, parasiticide and stimulent. Th eplant is indicated in abdominal enlargements, diabetes, diarrhoea, dyspepsia, epilepsy. Flatulence, insanity, leprosy, skin diseases-scabies, herpes and other cutaneous diseases; ulcers, fistulous sores, bleeding piles and whooping cough.

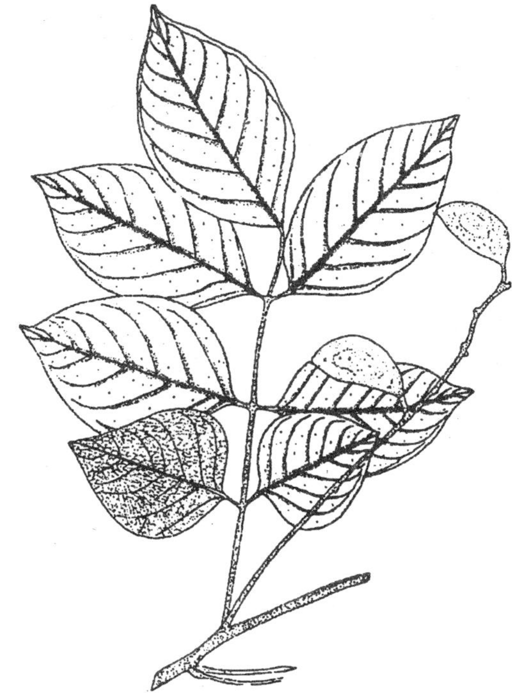

Pongamia pinnata

Populus angustifolia James.

Family : Salicaceae
Trade/Common Name : Narrow Leaf Poplar

A pyramidal shaped tree, upto 20 m height. Leaves having wavy margins and narrow along with scaly or often resinous buds.

It is found in various dry regions of India. It is propagated by seeds.

Flowers and ripe fruits have medicinal properties. The decoction of fruits and flowers are used to treat infected gum, toothache. To stop internal haemorrhage, a decoction of fresh flower is administered. Tea of fresh flower is used for drinking to purity the blood.

Populus balsamifera Linn.

Family : Salicaceae
Trade/Common Name : Balsam Poplar

A beautiful tall tree upto 35 m tall. Leaves are pale green below and shiny dark green above, upto 15 cm long and 10 cm wide, finely toothed, rounded or heart-shaped at the base. The buds are aromatic.

It is found in various dry regions of India. It is propagated by seeds.

Mature leaves, bark, flowers and flower bud are medicinal importance. The bud and fresh flowers are steeped in cold water and stained off, the liquid is drunk for purification of blood. To treat snake bite, the leaves and bark are boiled in water and vapour inhaled. For treating pulmonary ailments, kidney disorders, stomach disorder, scurvey, rheumatism and gout the tincture prepared from the tree bark is used.

Portulaca oleracea Linn.

Family : Portulacaceae
Trade/Common Name : Common Purslane

An annual succulent herb. Stem prostrate, rooting nodes, purplish green, leafly all over with axillary, yellow flowers and blooming in the month of June-July.

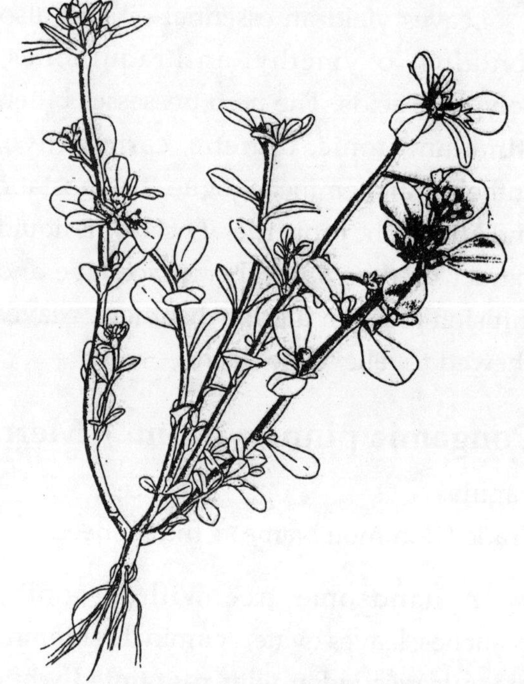

Portulaca oleracea

It is found in orchards, crop fields, kitch gardens, moist waysides, margins of crop fields throughout India commonly in temperate and sub-tropical Himalya.

The herb possesses refrigerant, anti-scorbutic, aperient and diuretic properties. It is useful in scurvy and diseases of liver, spleen, kidney and bladder. It also prescribed in the treatment of cardiac-vascular disorder, dysuria, haematuria, gonorrhoea, dysentery, sore nipples and ulcer of mouth. It is used as a blood purifier.

Potentilla fulgen Wall.

Family : Rosaceae
Trade/Common Name : Bajradanti

An erect, perennial, stout, ascending herb, 15-75 cm tall with a thick rootstock.

Potentilla fulgen

Leaves are pinnate, clothed with silvery hairs beneath. Flowers are orange yellow in corymbose panicles and blooming in the month of August to continue upto September.

It is found in temperate Himalaya between 2000-3000 m elevation. It is propagated by seeds and root slites.

The root is used to strengthen gums and teeth.

Prangos pabularia Lindley.

Family : Apiaceae
Trade/Common Name : Stony Herb

A perennial herb, bushy branched, leafy upto 1 m. Leaves long, many times divided into linear filiform segments. Flowers are yellow in colour and arranged in compound umbel and blooming in the month of July-September.

Prangos pabularia

It is found in stony slopes and dried places in western Himalaya. It is propagated by seeds.

Roots contain coumarins. Fresh leaves yield an essential oil (2%). It contains myrcene (48%), alpha-pinene (4%), resinous residue (28%). The herb possesses stimulant, carminative and diuretic properties. The roots are used in external application in treatment of itches, infusion given for indigestion and irregularity in menses. The fruit is promoted expulsion of faetus; used to cure foot rot in sheep.

Princepia utilis Royle

Family : Rosaceae
Trade/Common Name : Bhekhal

A thorny shrub upto 3.5 m in height. Leaves are lanceolate or elliptic-lanceolate. Flowers yellowish-white in axillary racemes and blooming in the month of December to continue upto March.

Princepia utilis

Globally, it distributed in China, India, Bangladesh and Farmosa. In India, It is found throughout Himalayan range from West to East between 1600-3000 m elevation. It is propagated by seeds.

Seeds yield semi-drying, pale yellow fatty oil (35-40%). It also contain resin (1.4%) The oil is used for cooking purposes and is an illuminant in the hills. The oil possesses rufefacient properties and is applied externally in the rheumatism and pains resulting from over fatigue.

Prunella vulgaris Linn.

Family : Lamiaceae

Trade/Common Name : Self-heal

A annual herb with many, spreading or ascending stems, 10-30 cm tall. Leaves opposite, ovate, obtuse and hairy, arising immediately below the bract and spreading beyond the inflorescence. Flowers are bluish, pick, purplish in terminal spikes and blooming in the month of November to continue upto February.

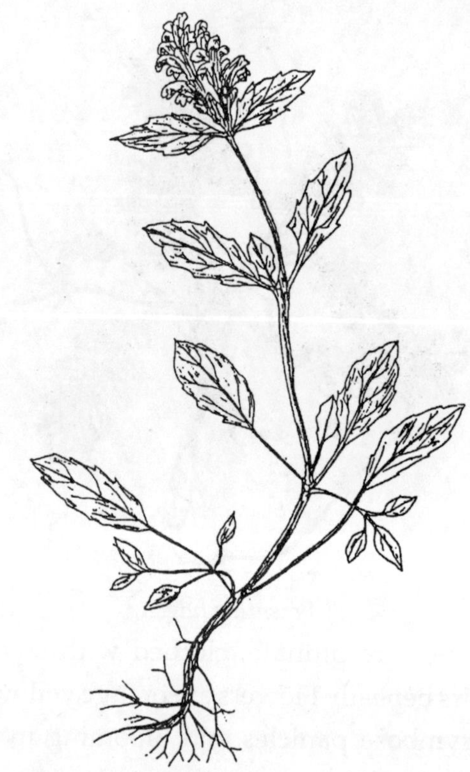

Prunella vulgaris

It occurs in open slopes, meadows, forest clearings, streamsides in forest from Afghanistan to Bhutan in the North temperate zone at an altitude of 1500-3500 m. It is propagated by seeds.

The herb contains a bitter principles and essential oil. The oil mainly consists of d-camphor, d-fenchone and traces of penchyl alcohol. The herb possesses antiseptic, expectorant, anti-rheumatism, anti-spasmodic and stimulant properties. It is used in fever and coughs, haemorrhages, diarrhoea, relaxed throat and ulcerated mouth in the form of an infusion and is useful for the treatment of internal bleeding piles.

Prunus cerasoides D. Don.

Syn. *P. puddum* Roxb. ex Wall.

Family : Rosaceae
Trade/Common Name : Paddam

A small or middle sized attractive tree. Bark is smooth, brown in colour. Leaves variable in length and breadth, ovate-lanceolate, dentate. Flowers are in fascicles or umbels and rose-red to white in colour and blooming in the month of November to continue upto February.

Prunus cerasoides

It is indigenous to the temperate Himalayan region, extending from Kashmir to Bhutan and in Akai & Khasi hills of Assam and in Manipur at altitudes of 900-2200 m. It is propagated by seeds, roots suckers and cuttings.

The stem bark contains flavonone sakuranetin, flavone genkwanin, isoflavone prunetin, isoflavonone and padmakastin. It possesses an anti-abortifacient, analgesic, carminative, conceptive, expectorant, fabrifuge and tonic properties. It is useful in stomach trouble, burning sensation of the body, cold & cough, seminal weakness and in pregnancy. Bark is mainly used for tannin.

Psoralea corylifolia Linn

Family : Fabaceae
Trade/Common Name : Babchi

A much branched annual heb upto 1.5 m in height. Stem and branches are covered with conspicuous glands and white hairs. Leaves round, dotted with black glands on both surfaces. Flowers are small, bluish purple, arranged on axils of leaves in branches and blooming in the month of September-October.

It is native to India and Iraq. It is found as weed in waste places in Madhya Pradesh. Chattisgarh, Uttar Pradesh, Uttaranchal, Rajasthan, Bihar, Jharkhand, Gujrat and Andhra Pradesh. It is propagated by seeds.

The pericarp of seeds contain psoralen, isopsorslen. It is used in indigenous system

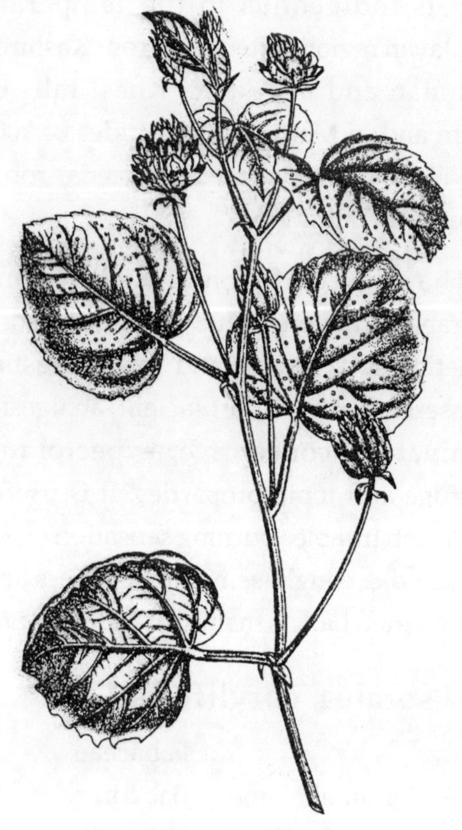

Psoralea corylifolia

Pterocarpus marsupium

of medicine in the treatment of leucoderma, leprosy and psoriasis. Besides psoralen, efficacy of psoralin is being investigated against several diseases including AIDS.

Pterocarpus marsupium Roxb.

Family : Fabaceae
Trade/Common Name : Indian Kino

A large tree. Leaves are compound with 5 to 7 leaflets. Leaflets are oblong or elliptic upto 12 cm long. Flowers are scented, yellow, upto 1.5 cm long, appear in very large dense bunches. Fruits are roundish, winged, to 5 cm long, each fruit possess one seed.

It occurs in Central and Penisular India in mixed deciduous forest. It is propagated by seeds or divisions.

Leaves, flowers and gum constitute the drug. It is useful in treating diabetic patients. Bruised leaves are applied on skin diseases, sores and boils. Its gum is astrigent and used to cure diarrhoea, fever and urinary discharges.

Pueraria tuberosa DC.

Family : Fabaceae
Trade/Common Name : Bankurma

A large climber, twining shrub with tuberous roots. Stem woody and leaves are trifoliate. Leaflets broadly ovate, acuminate, the lateral oblique and smaller size. Flowers blue or purplish blue in simple or panicled,

Pueraria tuberosa

terminal racemes and blooming in the month of May to continue upto June.

It is distributed throughout India, except in very humid or very arid regions and ascending upto 1200 m. It is propagated by seeds and crowns.

Tubers contain Vitamin B_1, sucrose, glucose and fructose. It possesses aphrodisiac, cardiac, emetic, expectorant and febrifuge properties. It is used in heart weakness and rheumatism. The roots are demulcent and refrigerant and are used in cataplasm on swollen joints and as a lactagogue.

Punica granatum Linn.

Family : Punicaceae
Trade/Common Name : Pomegranate

A large deciduous shrub or small tree. Bark smooth, dark grey. Branchlets sometimes armed with small, axillary or terminal thorns. Leaves opposite and long. Flowers bright red, tubular and blooming in the month of May to continue upto July.

It is native of Iran, Afghanistan and Baluchistan. It is found in growing wild in the warm valleys and outer hills of the Himalayas between 900-1800 m and is cultivated throughout India. It is propagated by seeds.

The rind contains gallotannic acid (28%) together with a yellow colouring matter. Flower yield sitosterol, ursolic acid, maslinic acid, elegic and gallic acid. The edible

Punica granatum

portion of the fruit contains glucose, fructose, maltose, starch, carotene, thiamine, riboflavin, nicotinic acid, vitamin C, pectin and amino acid. It possesses anthelmintic, astringent, cardiac, cooling, expectorant properties. It is useful in brain affections, coughs and colds, diarrhoea and dysentery, heart tonic, increases seminal flow and stop bleeding from the nose. Fruit is a good source of sugar and Vitamin C. Rind is used as an astringent in diarrhoea and dysentery. Flower –buds are used in bronchitis.

Putranjiva roxburghii Wall.

Family : Euphorbiaceae
Trade/Common Name : Putranjiva

A dioecious, evergreen tree, attaining a height of upto 18 m and a girth upto 2 m having gray bark. Leaves long, elliptic-oblong to ovate-lanceolate, unequal sides at the base, dark green in colour and shining in appearance. Flowers are small, male in dense rounded clustered, yellowish in colour, female flowers solitary or 2-3 together, green in colour and blooming in the month of March–April.

It is found wild or cultivated almost in all parts of India, ascending of 750 m. It is propagated by seeds.

The leaves and stones contain glucoputranjivin, glucocochlearin, glucojiaputin, gluco-cleomin, minnitol, saponin, glucosides and fatty oil. Its decoction is used for cold & cough, rheumatism. It is reported to have the properties. To effect conception in sterile omen and attributed with the birth of a male child, hence the name 'Putranjiva' or 'Jiyapota'.

Putranjiva roxburghii

Pyracantha crenulata (D. Don.) M. Roemer

Family : Rosaceae
Trade/Common Name : Choota seb

A spiny, evergreen shrub with usually crowded, narrow-oblong, shining, leathery leaves and numerous small white flowers in axillary clusters arranged along the branches. Leaves usually 1.5-3 cm; with obscure rounded teeth, spines stout, terminating lateral branches. Flowers are blooming in the month of April to continue May.

It is found Sutlej valley to Bhutan between 1000-2600 m. It is propagated by seeds.

Leaves contain leucocyanidins flavonoids princle-vitexin-4-rhamnoside, vitexin glycoside and oligomeric leucoantho-cyanidin. Fruits contain flavonols-kaempferol, quercetin and saponins. Fruits are credited with cardiotonic, coronary vasodilator and hypotensive properties. It is used for cardiac failure, myocardial weakness, paroxysmal tachycardia, hypertension, arteriosclerosis and Burgor's diseases.

Quercus leucotrichophora Cam.

Syn. *Q. incana* Roxb.

Family : Fagaceae
Trade/Common Name : White Oak (Banj)

An evergreen tree of 40 m tall with pale-grey to blackish, rough. Leaves ovate-lanceolate, acuminate, cuspidate-serrate, coriaceous, glossy dark green above, densely white pubescent beneath. Flowers solitary or clustered on short spikes, axillary and blooming in the month of March-April.

It is found abundantly on North-East slopes of Submontane ot montane Himalaya

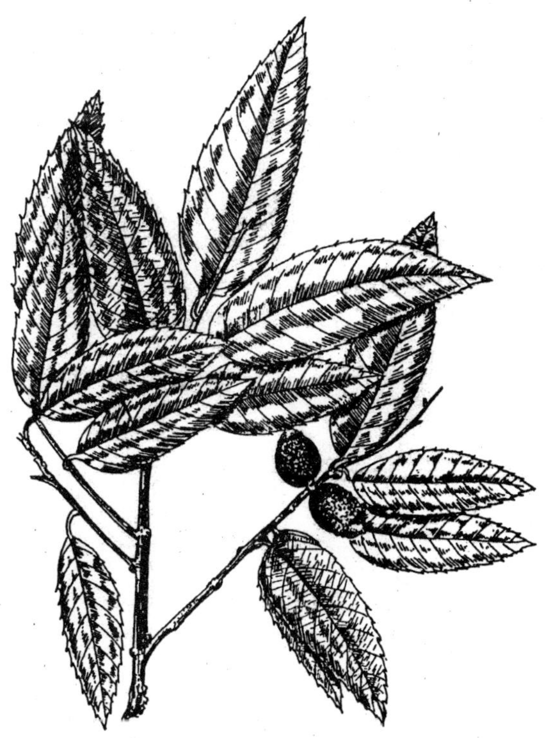

Quercus leucotrichophora

of Garhwal regions or otherwise usually associated with *Rhododendron arboreum* and *Myrica esculenta*. 800-2000 m altitude. It is propagated by seeds or divisions.

It is diuretic and used to treat gonorrhoea and diarrhoea.

Quercus semecarpifolia Smith.

Family : Fagaceae
Trade/Common Name : Brown Oak

An evergreen tree of 40 m tall with dark-grey cracked into scales. Leaves ovate-elliptic or oblong, often with entire margins on old trees, spiny teeth in young trees, obtuse, coriaceous, dark-glossy green above, rusty-pale tomentose beneath. Flowers

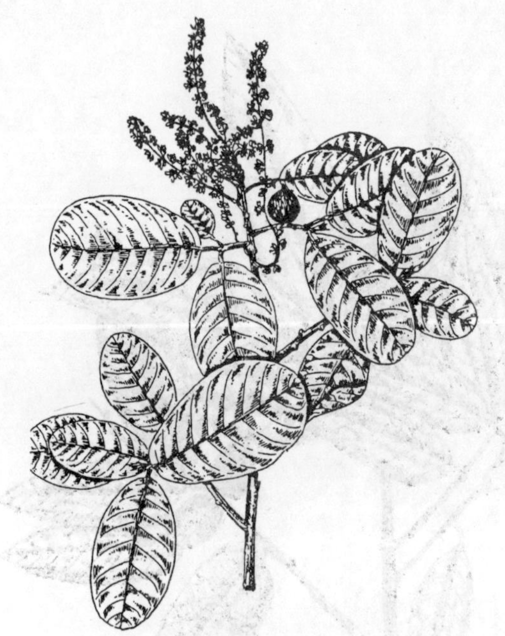

Quercus semecarpifolia

solitary or clustered on short spikes, axillary and blooming in the month of March-April.

It is found along the montane forests especially on north slopes of Garhwal to North-East states and associated with Silver fir and Deodar trees. 800-2000 m altitude. It is propagated by seeds or divisions.

It is diuretic and used to treat gonorrhoea and diarrhoea.

Randia dumetorum (Ratz.) Poir.

Family : Rubiaceae
Trade/Common Name : Madanphal

A large deciduous, thorny shrub or a small tree, with grey bark, often long and stout axillary spines. Leaves long, ovovate, obtuse or acute, narrowed into the short, marginate stipules. Flowers are greenish or whitish-yellow in colour, fragrant, solitary or 2-3 together at the ends of short, lateral branchlets and blooming in the month of December-January.

It is found throughout India an elevation of 1300 m in the dry regions of hills. It is propagated by seeds.

Bark, root and fruits are medicinally important. It contains saponin, valerianic acids, wax, resin, colourful matter, d-mannitol and essential oil. It possesses nauseant, expectorant, diaphoretic, anthalmintic and abortifacient properties. It

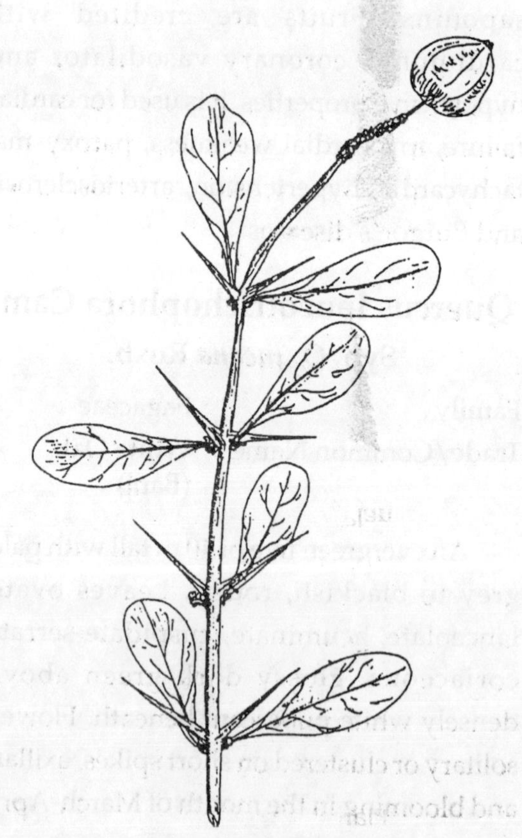

Randia dumetorum

is useful as a nervine, calmative and antispasmodic remedy for the teething ailments of the soap industry. Bark is astringent and is given in diarrhoea and dysentery.

Rauvolfia serpentina Benth. ex Kurz.

Family : Apocynaceae
Trade/Common Name : Sarpgandha

An erect, glaborous, undershrub, 30-75 cm high, Leaves are whorled, 8-20 cm long, gradually tapering in a short petiole. Flowers are white or pinkish, about 1.5 cm long in small clusters and blooming in the month of May to continue upto August.

It is found in almost all parts of India, upto an altitude of about 1000 m. It is more common in submountane regions of Himalayas and in lower ranges of Eastern and Western Ghats. It is propagated by seeds.

Rauvolfia serpentina

The root yields 0.8-1.3% total alkaloids, consisting of serpentine, reserpine, ajmaline, ajmalinine and ajmalicine. The chief of the drug is as a sedative and hypnotic and for reducing blood pressure. The drug is now largely used in insanity and high blood pressure. It is more suitable for cases of mild anxiety or patients of chronic mental illness. Root is also considered as anthelmintic and an antidote to snake venom.

Rheum australe D. Don.

Syn. *R. emodi* Wall. ex Meis.

Family : Polygonaceae
Trade/Common Name : Rhubarb

A robust perennial herb with thick, long roots. Roots are stout and the stem is hollow. Lower leaves are 20-50 cm in diameter. Orbicular and are broadly ovate. Flowers are small and dark purple in colour and blooming in the month of June-July.

The herb is distributed in the rocky lands, open slpoes, forest clearings, grsslands of temperate to alpine zone of Himalayas from Kashmir to Sikkim at altitudes of 3300-5200 m. It is propagated by rhizome cutting and seeds.

Rhizomes/roots yield an essential oil (0.05%) containing eugenol. It also contains chrysophanic acid, emodin, starch, rhein, anthraquinone derivated. It is purgative,

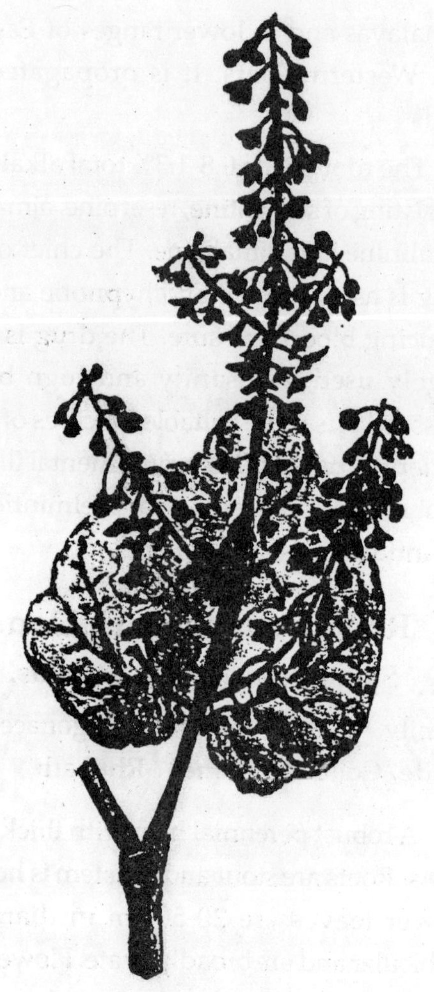
Rheum australe

astringent, tonic, stomachic and aperient in nature. Powered roots are used for cleaning teeth and are sprinked over ulcers for quick healing. Locally, the roots used as home remedies in stomach problems, cuts, wound and muscular swelling, tonsillitis and mumps.

Rheum moorcroftianum Royle.

Family : Polygonaceae
Trade/Common Name : Rhubarb

A stout, perennial herb variable sized. Stem lacking with radial leaves 10-60 cm diameter, thick, cariaceous, margins undulate, usually cordate often puberulous beneath. Flowers are pale-yellowish arising in stout, glabous racemes from middle of the large, stout radical leaves and blooming in the month of June to continue upto August.

It is found at altitudes of 3000-5200 m in Himalayas in Uttarnchal and Himachal Pradesh. It is propagated by rhizome cuttings.

It contains chief constituents, certain anthraquinone derivatives, glucoside rhaponticin, chrysophanic acid, emodin, eugenol and drug is used as purgative. Due to its astringent action, the drug is also given in certain types of diarrhoea.

Rheum moorcroftianum

Rhododendron anthopogon D. Don.

Family : Ericaceae
Trade/Common Name : Dhoop

A small shrub with aromatic odour. Branchlets are scabrous and scaly. Leaves are 2.5-4 cm long, elliptic or broad-oblong, shining above and brown tomentose beneath having a short petiole. Flowers are yellow and in dense, terminal corymbs and blooming in the month of June-July.

It is found in the Himalaya from Kashmir to Bhutan at altitudes of 3000-5200 m. It is propagated by seeds, divisions of sucker and layering.

Rhododendron anthopogon

Leaves possess stimulant properties. These are aromatic and are administered as an errhine to produce sneezing.

Rhododendron arboreum Sm.

Family : Ericaceae
Trade/Common Name : Bruans

An evergreen, much branched tree, upto 14 m in height and 2.4 m in girth. Bark is reddish-brown, soft, rough, exfoliates in thin flakes. Leaves are lanceolate or oblong, crowed towards the ends of the branches. Flowers are showy, red, in dense, globose cymes and blooming in the month of March-April.

The plant is found in the Himalayas from Kashmir to Bhutan and in the hills of Assam and Manipur at altitudes of 1200-4000 m. It is propagated by seeds and cutting.

Rhododendron arboreum

Green leaves contain a gluçoside, ericolin (arbutin), ursolic and á-amyrin, epifriedelinol, quercetin and hyperoside. Bark contains leuco-perlargonidin. Flower contains ursolic acid, quercitrin and resin. Young leaves are poisonous; applied to forehead for headache. Flowers are used in diarrhoea and dysentery.

Rhododendron campanulatum D. Don.

Family : Ericaceae
Trade/Common Name : Kashmir-patha

An evergreen, rhambling shrub or a small tree. Bark is thin, cinnamon-coloured or gray, smooth and peeling off in thin papery flakes. Leaves are elliptic or ovate, leathery and upper surface with rusty brown tomentosa. Flowers are white, in various shades of purple or pink, spotted within in corymbose heads and blooming in the month of May to continue upto July.

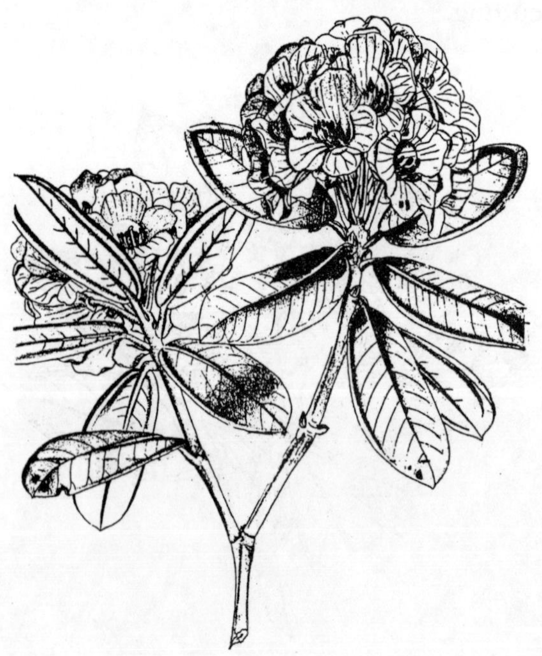

Rhododendron campanulatum

It is found throughout the Himalayas at altitude of 2400-5200 m. It is propagated by seeds and suckers.

Leaves contain a tonic substance, andromedotoxin, besides, ericolin, ursolic acid and a bitter, yellowish-brown resin. It is used in chronic rheumatism, syphilis and sciatica. They are mixed with tobacco and used as stuff to cure hemicrania and colds.

Rhynchostylis retusa Bl.

Family : Orchidaceae
Trade/Common Name : Cat Tail Orchid

An epiphytic herb. Stem are stout and woody, clothed with shealths of fallen leaves, upto 60 cm long. Leaves are leathery, strap shaped, linear, deeply channeled, gracefully arching, upto 30 cm long and 2.5 cm broad. Infloresences are 1 cm across, fragrant, long lasting, white, spotted with bluish purple.

It occurs in Jharkhand, West Bengal, Orissa, Western Ghats and Malabar Coast. It is propagated by bulbs.

Leaves are used to treat rheumatic diseases.

Ribes orientale Des.

Family : Grossulariaceae
Trade/Common Name : Gooseberry

A deciduous shrub, upto 4 m height with reddish brown branches. Leaves

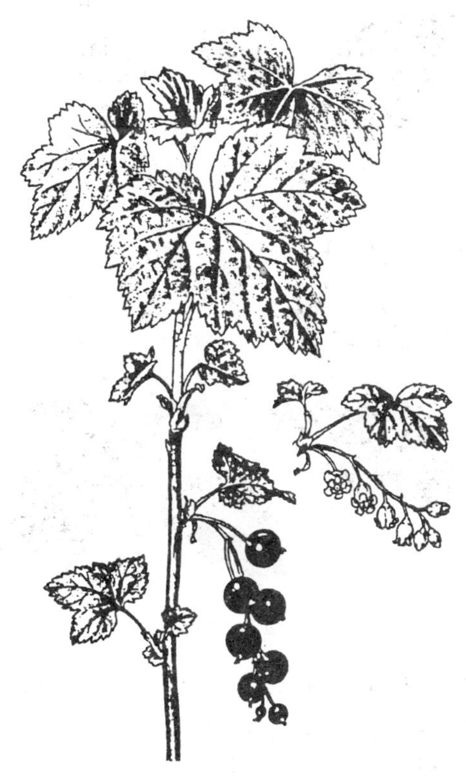

Ribes orientale

orbicular or reniform, stout, pubescent, glandular. Flowers purplish-brown, long racemes and blooming in the month of March to June.

It is found rarely in open rocky slopes of Montane and Alpine Himalayas from Kashmir to Bhutan, above 2700 m altitude. It is propagated by seeds or divisions.

Leaves decoction used as diuretic. The fruit is recommended for gastric trouble, spasmodic vomiting and dysentery.

Ricinius communis Linn.

Family : Euphorbiaceae
Trade/Common Name : Castor-bean

A tall, evergreen, soft-wooded shrub. Leaves are very large, broad, roundish in outline, but partly divided into 5-7 palmate lobes with toothed margins. Flowers are large in terminal branches and blooming in the month of May to continue upto July.

It is cultivated throughout India and also found wild in the fields, gardens, village thickets, along roadsides, near habitations and wastelands in the hills. It is cultivated in most tropical and warm temperate countries upto an elevation of 2000 m. It is propagated by seeds.

Seeds contain a fixed oil (45-50%), known as castrol oil. It contain a crystalline principle ricinine, the poisonous taxalbumin ricin. It is used as purgative. The use of castor oil for facilitating child birth is doubtful, rather it should be sparingly used even as purgative for pregnant women and during menses. The decoction has also been reputed to act as a lactagogue and emmenagogue, when administered internally.

Rosa brunonii Lindl.

Family : Rosaceae
Trade/Common Name : Himalayan Musk Rose

A sub deciduous, rambling climbers, 5-15 m tall with brown, rough bark. Leaves imparipinnate, rachis with prickle on the lower surface. Flower white in large axillary or terminal corymbs and blooming in the month of March-April.

It is found in open places, forest clearings, with shrubberies or Pine-Oak forest Edges, 600-2000 m altitude from Kashmir to Bhutan. It is propagated by grafting or cuttings.

Leaf and flower juice used in wound and ophthalmia. Dried flower-powder given in diarrhoea.

Rosa damascena Mill.

Family : Rosaceae
Trade/Common Name : Damask Rose

A perennial shrub attains 3 m height. Stem is covered with stout hooked falcate prickles of unequal and dilated towards the base. Leaf is stipulate, compound and imparipinnate. Leaflets are five to seven, ovate, rather rigid and oblong in bud and serrate. Flowers appear in groups of 5-7 and axillary terminal in corymbose. Flowers are sweet smelling and fragrance and blooming twice in a year during March-April and September-December.

It is native of the Orient. It is commercial cultivated in Aligarh, Etah, Ghazipur and Ballia Districts of U.P. and Udiapur district of Rajasthan. It is propagated by grafting or shoot cuttings.

Rose oil is extracted from rose petals. The aroma in flower is due to essential oil secreted in the papillae from epidermal cells. The rose concrete is extracted by volatile solvent (petroleum ether) which yield 0.25% flower concrete. The oil contains genaniol, citronellol, ehanol, rose oxide, linalool, nerol, eugenol etc. 'Rose water' is produced by hydro-distillation of rose flowers. More than

Rosa damascena

80% of rose flowers are used for the production of rose water in India. It is used for making syrups, medicinal preparation and also sprinkling in social functions. It is also used in eye lotion and eye drops for its soothing qualities. It is used in Unani medicine as good tonic, and mild laxative.

Rosa moschata Mill. ex. Herr.

Family : Rosaceae
Trade/Common Name : Musk Rose

A large thorny climber with reddish, sparsely prickly stems. Foliage dark green, with purplish flush and upto 4-7 leaflets, grayish below. Flowers are white, 5.0 cm across appear in terminal cluster, sweet scented. Hips ovate, orange red or dark brown.

It is found in Middle East, the Himalayas and Europe. In India, it occurs

Rosa moschata

in Uttaranchal Himalayas upto 2000 m altitude. It is propagated by grafting or cuttings.

Rose water prepared from flowers are used in medicine. It is also used in various perfumes.

Rosa sericea Linn.

Syn. *R. omeiensis* Linn

Family : Rosaceae
Trade/Common Name : Thorny Rose

A thorny sgrub upto 3 m tall. Leaflets ranging from 5-17 which look like those of fern. Flowers are white, single with four petals and 5 cm in diameter. Hips are small, pear-shaped and bright red.

It is distributed in Uttaranchal hills between 2400-3600 m altitude particularly at Kedar Nath, Badri Nath. It is also found

Rosa sericea

in North Bengal, Manipur, Sikkim, Bhutan and Assam.

Syrup prepared from flower is used to treat epilepsy and insomnia.

Roscoea alpina Royle.

Family : Zingiberaceae
Trade/Common Name : Safed-musli

A tuberous rooted, erect herb with oblong-lanceolate leaves. Tubers are fasiculated, spindle shaped, white, 2-5 cm long, arising in branches and Flowers are pink-purple in colour and blooming in the month of June to continue upto August.

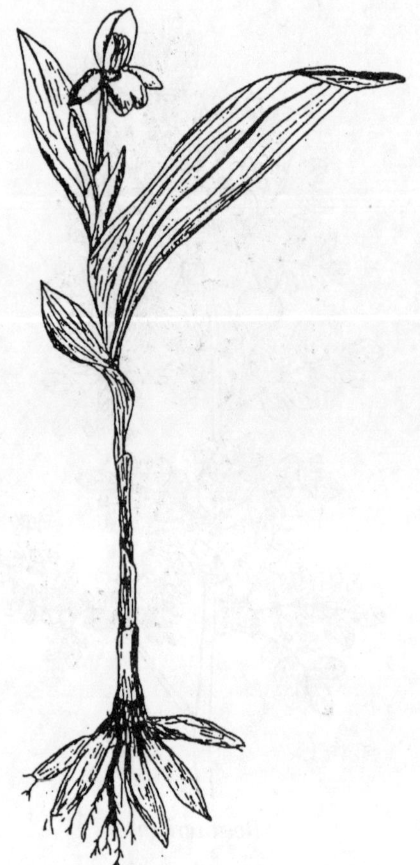

Roscoea alpina

It is found from Uttaranchal to South-East Tibet between 1500-3000 m elevation. It is propagated by tuberous roots.

Tuberous roots are useful as tonic in general debility and in given in impotency, diabetes, leucorrhoea, diarrhoea and dysentery.

Rosmarinus officinalis Linn.

Family : Apiaceae
Trade/Common Name : Rosemary

A hardy evergreen fragrant shrub of 2 m tall. Leaves are rolled back and stomata in hairy grooves on lower side. Flowers are violet blue in colour.

It is found Nilgiri hills and temperate Himalaya. It is propagated by cutting.

Flowers yield an essential oil which contains d-pinene, B-pinene, P-cymene etc. It is used as an stimulant, diuretic, aperient, emmenagogue. A balm is made from it for frictions, to stimulate the circulation and productive revulsive effects. Fresh tender

Rosmarinus officinalis

tops of these plants are used for garnishing and the for flavouring food, cold drinks, soaps etc. Dried leaves are used as condiments. The plant is astrient and it is considered as an excellent stomachic and a nervine tonic.

Rubia manjith Roxb. ex Fleming.

Family : Rubiaceae
Trade/Common Name : Indian madder

A prickly creeper or climbing, perennial herb upto 10 m long. Rootstocks are perennial and the roots are long, cylindric, flexuous with thin, red bark. Stems are 4 angled. Leaves are cordate-ovate to ovate-lanceolate, 2-8 in a whorl. Flowers are small, white or greenish, or in shades of red and yellow; sweet scented in many or few flowered terminal panicles or cymes and blooming in the month of June to continue upto September.

Rubia manjith

The plant is common throughout India, ascending upto an altitude of 3750 m. It is propagated by seeds and cutting.

Roots contain purpurin, munjistin, xanthopurpurein or purpuroxanthin, pseudo purpurin and alizarin. It possesses tonic, astringent, antidysentric, antiseptic and deobtruent properties. It is used in rheumatism and form an ingredient of several ayurvedic preparations. A decoction of the leaves and stems is used as a vermifuge.

Rumex dentatus Linn.

Family : Polygonaceae
Trade/Common Name : Jangli Palak

An annual, erect, glabrous herb upto 1 m tall with reddish roots and ribbed hollow stem. Basal leaves in close whorls, much larger to cauline leaves, oblong, obtuse, base rounded or cordate, margins undulate. Flowers short stalked arranged in leafy or leafless verticillate panicles and blooming in February to May.

It is found in waste places, road sides of the outer submontane zones of Submontane Himalaya and adjacent plains and South India. It is propagated by seeds.

Rumex dentatus

Decoction of root is used externally to treat skin diseases. The poultice of root is applied to skin itch, ulcers and skin diseases.

Salix alba Linn.

Family : Salicaceae
Trade/Common Name : White Willow

A ornamental tree upto 15 m tall with spreading and drooping branches. Young branches are silky white in colour and with maturity they turn olive-brown. Leaves are glossy green above and silky white below, narrow, pointed with finely toothed edges upto 10 cm long. Flowers are catkins.

Salix alba

It is mostly cultivated in North-West Himalaya. It is propagated by seeds or cuttings.

Young leaves and flowers are medicinally important. Tea made from the leaves and bark is used in fever. The bark preparation is known as a tonic in convalescence and also used for treatment of menstrual irregularities and diarrhoea.

Salvia lanata Roxb.

Family : Lamiaceae
Trade/Common Name : Ghanyajhar

A perennial, white, woolly-tomentose herbs upto 30 cm high. Leaves mostly radical, sessile, oblanceolate. Flowers are blue, orbicular, glandular-hairy, abruptly pointed and blooming in the month of March to June.

It is found in open, exposed grassy slopes, 800 m and above in submontane to

Salvia lanata

montane of North West Himalaya to West Nepal. It is propagated by seeds or divisions.

Leaf infusion is used in colic and diarrhoea. Flower paste used in cold and cough.

Salvia moorcroftiana Wall. ex Benth.

Family : Lamiaceae
Trade/Common Name : Kalijari

A tall, robust, perennial 45-90 cm tall herb. Leaves are thick, long stalked, ovate or oblong, sinuately and irregularly lobed, hairy; lower surface white tomentose. Flowers are pale blue or white with red

Salvia moorcroftiana

coloured bracts arranged in distant whorls on a long spike and blooming in the month of March-April.

It is found in the patches in the temperate Himalayas from Kashmir to Uttaranchal upto 1800 m elevation. It is propagated by seeds.

Root yields an essential oil and mucilage. It is used in cold and coughs. Leaves are used for guinea-worm and are applied as poultice for boils, wounds and chronic affections of the skin. Seeds are emetic and are used in haemorrhoids, colic and dysentery.

Sambucus wightiana Wall. ex Wright & Arn.

Family : Sambucaceae
Trade/Common Name : Wild Bony Herb

A perennial herb upto 1.5 m tall. Leaves large pinnate, terminal flat-topped clusters of numerous tiny white or creamy flowers.

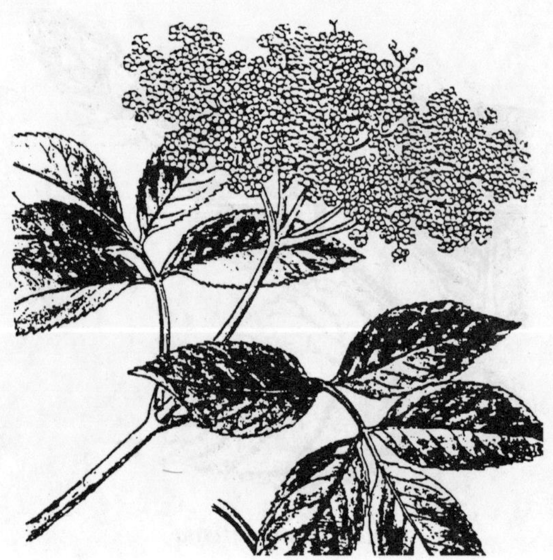

Sambucus wightiana

It is very common on heavily grazed slopes, forest floor, waysides, in gregarious clumps. It is propagated by seeds.

The herb is reported to be used for purgative properties. The decoction of root and inner bark of dwarf elder is an effective diuretic.

Santalum album Linn.

Family : Santalaceae
Trade/Common Name : Sandal Wood

A evergreen tree with drooping branches. Bark is dark, rough with vertical cracks, mature wood scented. Leaves are opposite, shining on upper surface upto 7 cm long. Flowers are small, purplish, appear in small bunches.

It is native to India. It is distributed in Deccan Peninsula. It is propagated by seeds or cuttings.

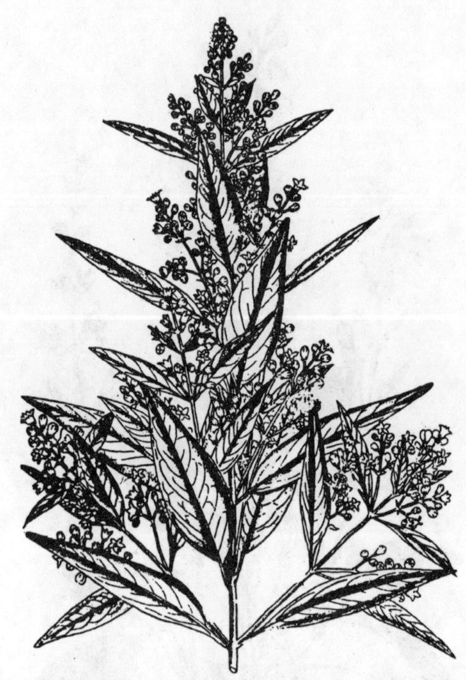

Santalum album

Heartwood yields an essential oil and is used to promote urination, inflammation of bladder, gonorrhoea and cough. Ethnobonatically, the wood, ground up with water into a paste and this paste is applied on skin diseases, local inflammations and on the forehead in fever. The oil is also used in perfumery industry.

Sapindus mukorossi Gaertn.

Family : Sapindaceae
Trade/Common Name : Soapnut

A very handsome tree green in summers and deciduous in winters, upto 18 m or more in height and 1-8 m in grith. Bark is dark-greenish or grey, or brown and is fairly smooth. Leaves are alternate, paripinnate; leaflets 5-10 pairs, glabrous, lanceolate,

Sapindus mukorossi

acuminate. Flowers are small, polygamous and mostly bisexual in terminal thyrses or in compound, cymose panicles and blooming in the month of May to continue upto June.

Tree is found in the Himalayas from Himachal Pradesh eastwards and in Assam ascending to an altitude of 1500 m. It is propagated by seeds.

Fruits contain saponin (7.2-13.8%) and mukorosside. It is used as expectorant, useful in slivation, chlorosis and epilepsy. It is a good substitute for washing soaps.

Saraca indica Linn

Syn. *S. asoca* (Roxb.) De Wilde

Family : Caesalpiniaceae
Trade/Common Name : Ashoka

Saraca indica

A medium sized evergreen tree, reaching upto 10 m with speading branches. Bark is dark brown. Leaves are alternate, compound upto 30 cm long with 6-12 lealets which are oblong lanceolate, acute, short petiolate. Young shoots are pendent. Flowers are yellow or orange and gradualy turn vermillion and scented at night.

It is widely distributed in West Coast of Bombay, in Khasi Hils, Chittagong, Arakan, Tenasserium, Upper Burma, Ceylon and Malaya. It is propagated by seeds.

Flower, seeds and dried bark are medicinally important. Powdered flowers mixed with water is given for haemorrhagic dysentery. Seeds are used to cure urinary discharge. Dried bark is astringent and used to treat excessive menstruations as a uterine sedative.

Sarcococca saligna (D. Don.) Muell. Arg.

Family : Buxaceae
Trade/Common Name : Sweet Box

A hardy evergreen shrub upto 1.5 m tall. Flowers are scentless, unisexual without petals, greenish in colour.

It is native to Northern India. It is found in commonly in Donpur regions (Uttar Kashi) of Uttaranchal. It is propagated by cuttings.

The root's decoction is useful in phosphoteuria and gonorrhoea.

Satyrium nepalense D. Don.

Family : Orchidaceae
Trade/Common Name : Satyrion

A terestial orchid upto 75 cm tall. Tubers are oval, producing stolon each with a small tuber at its end. Stem glabrous, sheathed near the base. Leaves are 2-3, mostly basal, fleshy, narrowly elliptic. Infloresences are densely many flowered, bracts are much longer than flowers. Flowers are fragrant, pink or white.

It occurs in Kashmir, Khasia hills, Darjeeling, Manipur and hilly regions of Southern India between 2700-3000 m altitude. It is propagated by seeds.

Roots are used to treat malaria and dysentery. It is considered as a tonic.

Saussurea costus (Falc.) Lipsch.

Syn. *S. lappa* (Decne.) Sch. Bip.

Family : Asteraceae
Trade/Common Name : Costus

A tall, perennial herb upto 2 m high. Leaves are very large at the base, borne on the winged stalks and upper leaves are smaller, sometimes with two lobes at the base of the leaves, almost clasping the stems. Flowers are about 2 cm long, bluish purple or almost black, borne on rounded flower heads and blooming in the month of June to continue upto August.

It is found in moist rocky slopes and moist ravines in forest from Kashmir to

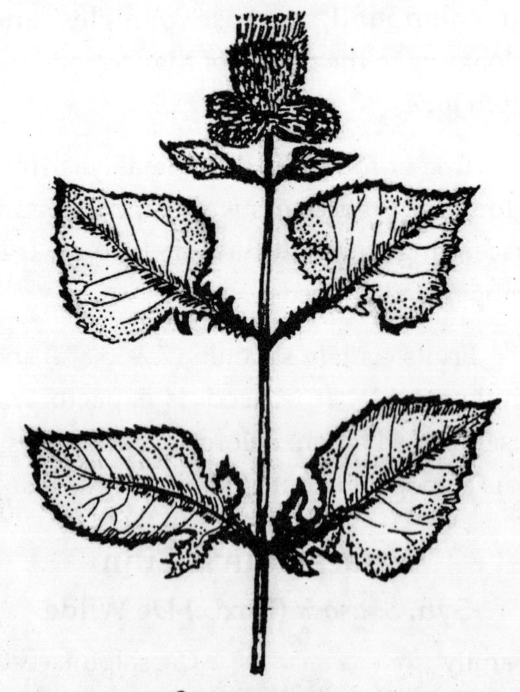

Saussurea costus

Uttaranchal at altitudes of 2600-4000 m. It is propagated by root cutting and seeds.

The root contains a volatile oil (1.5-2.5%), an alkaloid saussurine (0.05%), resin (6%) and tannins. It possesses anodyne, anti-arthritic, antiseptic, aphrodisiac, astringent, carminative, digestive, diaphoretic, deodorant, disinfectant, diuretic, emmenagogue, expectorant, febrifuge, narcotic, spasmodic, stimulant, stomachic, tonic and vermifuge properties. It is useful in asthma, bronchitis, colic, cough, dental trouble, diarrhoea and dysentery, fever flatulence, headache, hiccough, hysteria, in some cases of heart trouble, menstruation trouble, rheumatism and promotes urination.

Saussurea gossypiphora Don.

Family : Asteraceae
Trade/Common Name : Saussuree

A dwarf shrub upto 15 cm tall. Flower heads are in the form of brown spots at the top. From a distance the plant resembles a dove.

It is rarely occurring in glaciated rocky slopes in Thamsar jol (Bara Bhangal), Dainasar (Chhota Bhangal), Chansal Peak in Simla and Sangla Pass in Kinnaur districts of Himachal Pradesh. It is propagated by seeds or rhizome cuttings.

The wool of the herb is applied to fresh cut, where it sticks easily, seals the wound, check bleeding and enhances healing.

Saussurea obvallata (DC.) Edgew.

Family : Asteraceae
Trade/Common Name : Brahma Kamal

It is an erect tufted herb to 70 cm tall.

It occurs in Rohal Kanda, Chansal Peak and Chandramohan of Simla district, Sangla and Chitkul passes of Kinnaur district of Himachal Pradesh. It is propagated by seeds or rhizome cuttings.

The bracts are boiled in water and used as hot fomentation to cure hydrosele. The roots are often applied as a paste in cuts and bruises.

Saussurea obvallata

Schleichera oleosa (Lour.) Oken.

Family : Sapindaceae
Trade/Common Name : Ceylon Oak

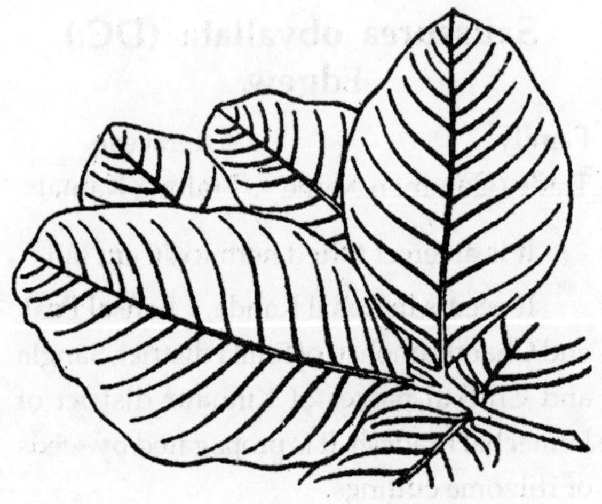

Schleichera oeolsa

A large tree. Leaves are red when young. Mature leaves are uniformly green. Flowers are greenish yellow in colour, often arises on short branches.

It is native to Indo-malayan regions. It is found in Geldubba of Chhindwara district of Madhya Pradesh. It is propagated by seeds.

Seeds yield an oil. It is used in leg swelling. The seed oil is used as illuminant, in perfumery and for soap making.

Selinum vaginatum C. B. Clarke.

Family : Apiaceae
Trade/Common Name : Bhutkeshi

A glabrous, stout herb, 1.5 m tall with robust, aromatic rootstock beset with thick tuft of grey fibres, left over by the dying leaves. Roots long, abruptly tapering from the middle. Stem is hollow and finely grooved. Leaves pinnate, seagents, lanceolate, serrate, lobed or pinnatifid. Flowers white, in long-stalked, compound umbels and blooming in the month of July to continue upto August.

It is founding temperate and alpine moist pastures in Western Himalaya between 1800-3800 m. It is propagated by seeds.

Roots contain coumarins like selinidin, vaginidin, angelicin, oroselol and lomatiom. Dried roots yields an aromatic oil (1.54%) containing β-pinene (45.5%), limonene (25.3%), camphene (5.2%). It possess a sweat, musky odour with a bitter taste and are employed as incense and used as nervine sedative. The oil has hypotensive, sedative and analgesic properties.

Selinum wallichianum (DC.) Rai. Sax.

Syn. *S. tenuifolium* Wall.

Family : Apiaceae
Trade/Common Name : Muramansi

An erect herb, 0.6-2.4 m tall with a perennial rootstock. Leaves-pinnately compound, deeply cut into narrow, linear, linear-lanceolate, acute segments. Flowers white in compound umbels and blooming in the month of July August.

It is found in temperate pasture lands, forest shades and alpine pastures from Kashmir to Uttaranchal at altitudes 2000-4000 m elevation. It is propagated by seeds.

The roots contain a volatile oil, angelicin, xanthotoxol and a mixture of long chain os saturated fatty oils. It is used as a nervine sedative and used as a substitute for Jatamansi. The volatile oil has hypotensive, sedative and analgesic properties.

Senecio jacquemontianus (Dec) Benth. ex Hook. f.

Family : Asteraceae
Trade/Common Name : Yellow Tagoma Herb

A perennial herb. Rootstalk stout, branched. Stem stout with broadly cordate basal leaves and oblong-obovate stem leaves. Flower heads with golden yellow ray-florats, in dense conical terminal clusters.

It is found alpine meadows, slopes, moist places in temperate western Himalaya at elevation 3000-4000 m. It is propagated by root cutting.

The roots are poisonous. Its poultice is applied to external boils or swellings for quick relief. Locally, water extract of leaves is a used for skin diseases.

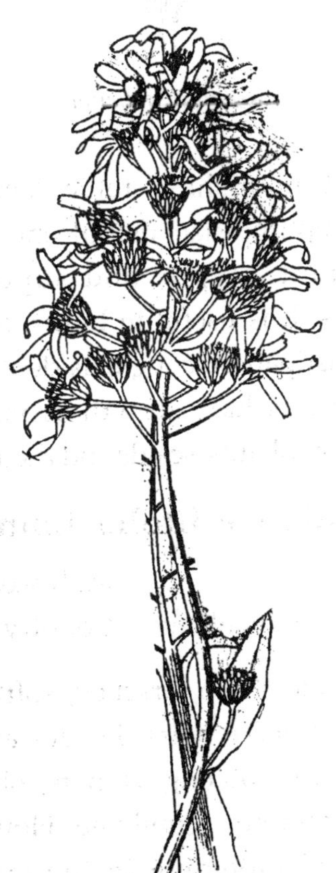

Senecio jacquemontianus

Senecio nudicaulis DC.

Family : Asteraceae
Trade/Common Name : Ragwort

Senecio nudicaulis

A erect, hardy, perennial, evergreen herb upto 80 cm tall. Flowers are bright yellow.

It is found in Kaream region of Chhindwara district, M.P. It is propagated by seeds or root cuttings.

An infusion of the herb in water is used to cure coughs, weakness due to prolonged fever and nervous disorder. A tea prepared from flowering tops is used to relive pains of child birth, to hasten decalyed child birth and also for kidney problems. It is used as a tonic and stimulant of the glandular system.

Sesamum indicum

Sesamum indicum Linn.

Family : Pedaliceae
Trade/Common Name : Sesame oil

An erect, much branched, strongly, smelling, hairly annual herb. Leaves large, thin sparsely hairy, upto 12 cm long, lower ones often lobed, toothed. Flowers axillary, solitary, whitish-pink having a offesive smell and blooming in the month of August to continue upto September.

It is a Kharif crop in most parts of India either as a single crop or mostly as intercrop with horsegram, buck-wheat and rainfed paddy.

Seeds yield a fatty oil. The leaves are highly mucilaginous and are used in the form of infusion as demulcent in infantile diarrhoea, cholera, dysentery, catarrh, bladder troubles, acute cystitis, strangury and respiratory troubles. The seeds possess emollient, nourishing, tonic, lactagogue, laxative, aphrodisiac, demulcent, diuretic, expectorant, styptic and suppurative properties. It is useful in burns, bleeding, piles, constipation, cough, as dentrifice, in dysentery, as hair growth promoter, in urinary complaints, scalds and rheumatism.

Sida cordifolia Linn.

Family : Malvaceae
Trade/Common Name : Country mallow

An undershrub, brached, softly, hairy and with stellate hairs. Leaves are long, cordate, ovate-oblong, crenate, obtuse or subacute and not acuminate. Flowers are yellow and blooming in the month of August-September.

It is a common weed distributed throughout tropical and sub-tropical India and Ceylon in waste places and scrub jungles upto an altitude of 1000 m. It is propagated by seeds.

The herb contains an alkaloid (0.09-0.3%), a fatty oil, phytosterol mucin and resin. It possess tonic and aphrodisiac properties. It is used for the fever, fits, leucorrhoea, micturition, gonorrhoea, colic, nervous disorders, general debility and heart irregularity. The bark is effective in curing facial paralysis and sciatica and the leaves are used for curing the bloody flux. Useful in ophthalmia, rheumatism and improves sexual strength.

Silybum marianum (Linn.) Gaertn.

Family : Asteraceae

Trade/Common Name : Holy-thistle

An erect, glabrous, thistle-like annual or biennial herb upto 1 m tall. Leaves large, variegated with spinose-toothed lobes, conspicuously white-spotted above. Flowers are rose purple in colour.

It is found from Western Himalaya, Kashmir and Punjab. It is propagated by seeds.

The seeds contains sylimarin and sylibin. Its seeds and leaves are used for their medicinal value. It is used against jaundice,

Sida cordifolia

Silybum marianum

inter-mittent fevers, dropsy and uterine trouble besides being reliable galactagogue. Seeds are pungent, demulscent and anti-spasmodic and used for the treatment of calculi of liver and gall bladder and in controlling haemorrhages.

Skimmia anquetitia Taylor & Airy Shaw

Syn. ***S. laureola*** **(Wall.) Hook.f. in part**

Family : Rutaceae
Trade/Common Name : Kastur Chara

A sttongly aromatic, evergreen shrubs with 1-1.5 m height. Leaves simple, crowed towards the ends of the branches, 8-17 cm long, oblong-elliptic or oblanceolate, entire, thick, gland dotted. Flowers yellow, in compact terminal panicles and blooming in the month of April-May.

It is found in temperate forest and shrubberies between 2400-4000 m elevation. It is propagated by seeds.

Leaves yield an essential oil, uses for high grade perfumes and cosmetics. Locally, the leaves are used as Haven Samagri and burn in temples to produce aroma.

Smilax aspera Linn.

Family : Smilacaceae
Trade/Common Name : Prickly Ivy

A climber shrub with thick, creeping rootstock. At the base of leaves two tendrils one on each side, usually regaded as stipules. Stem are furnished with recurved hooks. Flowers are white in colour.

It is native to Europe. It occurs in Garhwal hills of Uttaranchal upto 1750 m altitude. It is also distributed throughout the greater part of India from Kashmir, Sikkim, Assam and Deccan Plateau. It is propagated by seeds or cuttings.

Root contains steroidal sapogenins. It is uses as blood purifier and to treat skin diseases.

Solanum anguivi Lamk.

Syn. ***S. indicum*** **auct. non L.**

Family : Solanaceae
Trade/Common Name : Brahati

An erect woody undershrubs, upto 1.5 m high. Branches tomentose, armed with

Solanum anguivi

recurved prickles. Leaves ovate, sunentire, sinuate or lobes, stellately pubescent above, tomentose beneath, nerves prikly on both sides. Flowers blue, in extra-axillary racemose cymes and blooming in the month of July-August.

It is found in wastelands, open slopes along roadsides upto 1000 m elevation. It is propagated by seeds.

The herb possess cordial, aphrodisiac, astringent, carminative, cardiac, tonic and resolvent properties. It is used for the treatment of asthma, dry cough, chest pain, chset pain, chronic febrile affection, colic with flatulence, worms and dysuria. The root constitutes one of the 'Dasamool' of Ayurveda.

Solanum khasianum

Solanum khasianum C. B. Clarke.

Syn. *S. viarum* Dunal

Family : Solanaceae
Trade/Common Name : Kantkari

A stout, much-branched, under shrub, upto 1.5 m tall, with almost straight prickles. Leaves ovate, lobed and the lobes are lanceolate or triangular, hirsute and prickly on both the surfaces. Flowers are white in colour and are lateral, 1-4 flowered racemes and blooming in the month of May to continue August.

It is found in the waste places, ravines in the Khasi, Jaintia and Naga hills of Assam and Manipur, Nefa, Sikkim, West Bengal, Orissa, the upper Gangatic plains upto an altitude of 1600 m. It is propagated by seeds.

The berries contain solasodine (4%), solakhasianin and diosgenin. It is used in various formulations including birth control pills.

Solanum nigrum Linn.

Family : Solanaceae
Trade/Common Name : Makoi

An erect or rhambling, sparingly or often branched, usually glabrous herb. Leaves are ovate or oblong, sinuate-toothed or lobed, narrowed at both ends. Flowers are white in colour and blooming in the month of June to continue upto September.

Solanum nigrum

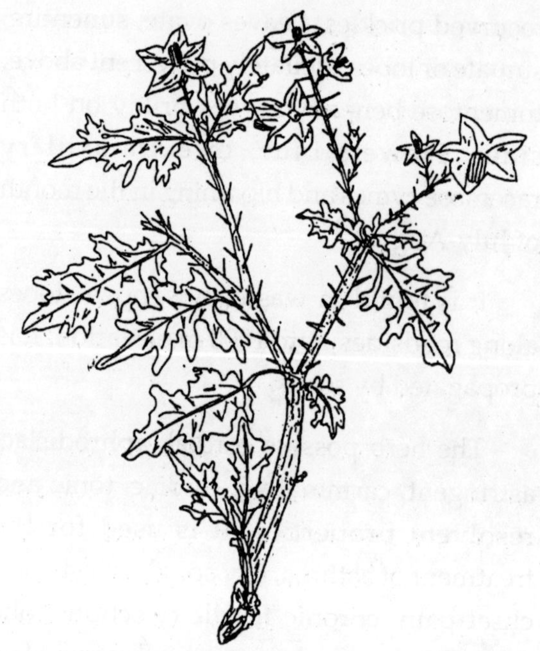

Solanum surattense

It is found as a weed near villages, footpaths and fields throughout India in dry parts, upto an elevation of 2100 m. It is propagated by seeds.

Leaves contain riboflavin, nicotinic acid and vitamin C. The immature green fruit of the plants contain four steroidal glyco-alkaloids viz. solamargine, solasonine and α-solanigrine. It possess antiseptic, emollient, diuretic, laxative, tonic, cathartic and antidysenteric properties. It is useful in wounds, sores, in treatment of psoriasis, eczema, piles and syphilis.

Solanum surattense Burm. f.

Syn. ***S. xanthocarpum*** **Schrad.**

Family : Solanaceae

Trade/Common Name : Wild egg plant

A prinkly, much branched herb, usually spearding or diffuse; young branched are densely covered with minute star-shaped hair, princkles are yellow, shining. Leaves are upto 10 cm long, their midribs and other leaves with sharp, yellow princkles. Flowers are purple, few together in small branched, opposite to the leaves and blooming in the month of April to continue upto July.

It is found throughout India, Ceylon and Pakistan. It also occurs in South-East Asia, Malaya and tropical Australia. It grows naturally in waste land, roadsides, dry places and field borders in warmers areas of country. It is propagated by seeds.

Fruits yield carpesteral, solasodine, solasonine and solanocarpine. The durg is useful in cough, asthma, pain in chest and fevers. It possess diuretic and antibacterial properties. It also useful in bronchitis, muscular pain, enlargement of liver, spleen, blood cancer, vomiting, and controlling of stone in bladders.

Solena amplexicaulis (Lam.) Gandhi

Syn. ***S. heterophylla*** **Lour.;** ***Zehneria umbellate*** **Thw.**

Family : Cucurbitaceae
Trade/Common Name : Ban-Kundru

A perennial, scandent or climbing herb; tendrils simple. Leaves much variable in shape and size, more or less 3-lobed or deeply 5-lobed, ovate-triangular to sagittate. Flowers are pale white, male flowers in subumbellate racemes; female flowers solitary or in pairs, axillary, short pedicelled and blooming in the month of July-August.

It is found in throughout India, commonly in climbing on bushes, forest edges, scrub jungles or nearly crop fields, from tarai-bhabar to 2000 m. It is propgated by seeds.

Ripe fruits are considered as stomachic and tonic. Dried powdered roots are given in the dose of 2 gm twice a day with milk in seminal debility and spermatorrhoea.

Sphaeranthus indicus Linn.

Family : Asteraceae
Trade/Common Name : East Indian Globe Thistle

Sphaeranthus indicus

A hairy herb, glandular with many branches, upto 75 cm tall. Flowers are pink or purple in colour.

It occurs at Chhindwara district of M.P. It is also found in the moist damp places of tropical zones of Garhwal Himalaya. It is propagated by seeds or cuttings.

The herb contains an essential oil (0.22%) and possess bitter alkaloid. It is aued as anthelmintic, aphrodisiac and stomachic. Locally, its stem, leaves and flower paste are used six to seven times a day for the fortnight in small pox.

Spilanthes acmella Linn.

Family : Asteraceae
Trade/Common Name : Akarkara

An annual herb, more or less pubescent, sometimes hairy. Stem 30-60 cm, usually

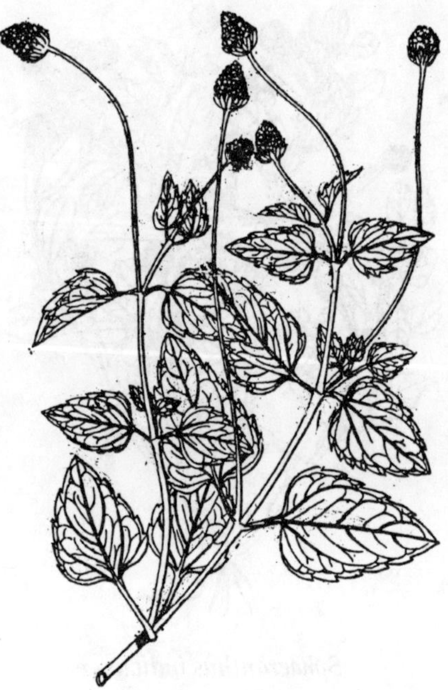

Spilanthes acmella

decumbent near the base. Leaves opposite, stalked, ovate-lanceolate, toothed or entire. Flower heads discoid or radiate, conical, solitary on long stalks. Flowers are white or yellow in colour, mostly bisexual and blooming in the month of May to continue upto August.

It occurs throughout India, ascending to 1700 m, often found occurring as a weed in rice fields. It is propagated by seeds.

The herb contains spilanthol. Leaves are chewed to relieve toothache and affections of throat and gums. Tincture made from flower heads to treat inflammation of Jaw-bones and caries. It increases the flow of saliva and is useful in fever especially during summers.

Spiraea tomentosa Linn.

Family : Rosaceae
Trade/Common Name : Hardhack

A shrub of 1.2 m tall. Leaves are toothed, woolly beneath, yellow tomentosa, up to 7.5 cm long. The branches are woolly and reddish brown in colour. Flowers are pinkish or white in colour, borne in spike like racemes.

It is native to North Temperate region and Himalayas. It is propagated by seeds or cuttings.

Leaves, flowers, stems and root bark of this plant are medicinal useful. A tea made from leaves and bark is used to treat diarrhoea. Tea made from flowers is given to pregnant women to easy child birth.

Stellaria media (L.) Vill.

Family : Caryophyllaceae
Trade/Common Name : Chick Weed.

A hardy, annual herb upto 30 cm tall. Leaves are ovate in shape. Flowers are white.

It is found throughout the temperate regions of India. It is propagated seeds.

Plant in full bloom are useful. It is used for eye infection and haemorrhoids. A poultice made from the plant is used to treat ulcers and skin sores.

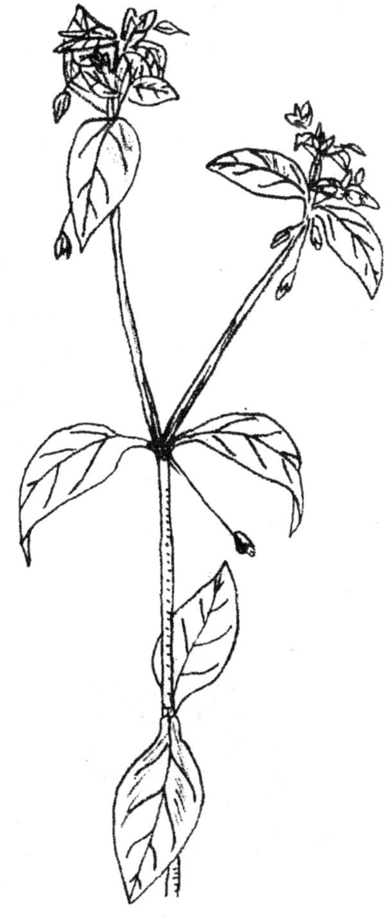

Stellaria media

Sterculia urens Roxb.

Family : Sterculiaceae
Trade/Common Name : Gum Karaya

A tropical medium sized deciduous tree upto 10 m tall with erect, straight trunk and broad top. Bark is smooth, glossy, papery, grayish-white, easily peeled off. Leaves are crowded at the end of branches, alternate, palmate, shallowly, fine lobed, very densely hairy on the under surface. Flowers appear in large erect terminal panicles, yellow brown in colour.

Sterculia urens

It is native to tropical Himalayas and Southern India. It is found in various areas of Central India. It is propagated by seeds or cuttings.

Gum is polysaccharide in nature and is used in throat infection and in dental fixture powders. It is also used in toiletries and cosmetics.

Sterculia villosa Roxb.

Family : Sterculiaceae
Trade/Common Name : Oadal Tree

A deciduous tree with grey bark. Flowers appear in bunches, yellow in colour. Fruits are follicular and seeds are black.

It is found throughout India. It is propagated by seeds or cuttings.

Sterculia villosa

The white gum exudes is used to treat throat infection and in dental fixture powders, lozenges and pastes.

Strychnos nux vomica Linn.

Syn. *S. colubrine* Wight

Family : Strychnaceae
Trade/Common Name : Nux vomica

A middle sized deciduous tree up to 15 m tall, often with short axillary spines. Leaves elliptic-orbicular, cariaceous. Flowers greenish-white, fragrant, in terminal compound cymes and blooming in the month of March-June.

Indigenous to India but growing in Burma, China and Australia. It is propagated by seeds.

Fruit contains glucoside loganin. Leaves and root bark contain brucine, strychnine, vomicine and methoxystrychnine. It is highly toxic to man and animals, producing stiffness of muscles and convulsions. Root is bitter, useful in intermittent fevers. Root bark, ground up into fine paste with lime juice and made into pills are used for cholera. Leaves are applied as poultice to sloughing wounds and ulcers, specially in cases when maggots have formed. Seeds are bitter, tonic, nervine tonic, stomachic, antidiarrhoeal, antidysenteric, antispasmodic, emetic, febrifuge, spinal, respiratory and cardiac, stimulant, aphrodisiac.

Strychnos nux vomica

Swertia chirayita (Roxb. ex. Flem) Karst.

Syn. *S. chirata* **Buch. Ham.**

Family : Gentiaceae
Trade/Common Name : Chirayita

An erect, annual, 60-125 cm tall. Stem robust, branching, cyclindrical below, 4 angled uptorwds and contain a large pith. Leaves broadly lanceolate, 5 nerved, sub-sessile. Flowers are lurid greenish yellow, tinged with purple, arranged on panicles and blooming in the month of July to continue upto September.

The plant is found in the temperate Himalayas at altitude of 1200-3000 m, from Kashmir to Bhutan and in the Khasi hills in Meghalaya at 1200-1500 m. It is propagated by seeds.

Swertia chirayita

The entire plant con tains ophelic acid (yellow and bitter), two bitter glucosides – chiratin and amarogentin, gentiopicrin, two yellow crystalline phenols and a new xanthone etc. It is used as tonic and bitter without aroma or astringency. It tends to produce a regular action and causes a free discharge of bile. It is useful for asthma, liver disorder, dyspepsia. It is credited with tonic, anthelmintic, anti-diarrhoeal properties in Indian system of medicine.

Swertia ciliata Buch. Ham.

Family : Gentiaceae
Trade/Common Name : Kala Chirata

A perennial herb. The corolla segment each bear two nectarines on th eupper side, consisting of little pits covered with hairs.

It is native to Himalaya and commonly occurring throughout temperate and alpine slopes of Centeral Himalaya near the forest edges and moist slopes upto 4000 m altitude. It is propagated by seeds or cuttings.

Leaves are used for medicinal purposes for fever.

Symplocos paniculata Linn.

Syn. *S. crataegoides* **Buch-Ham. ex D. Don.**

Family : Styraceae
Trade/Common Name : Lojh

A deciduous shrub or a tree, upto 12 m in height. The bark is soft, friable, of light brown colour. The external surface is corky or much fissured transversely; while internal part is of a lighter colour and of fibrous texture. Taste is faintly balsamic and astringent. Flowers are white, fragrant, in terminal panicles and is blooming in the month of May to continue upto July.

The tree is common in the Himalayas from Himachal Pradesh to Assam and in the Khasi hills, occurring at altitudes between 600-2400 m. It is propagated by seeds.

The bark contains two alkaloids; loturine and colloturine and also a red colouring matter. It occurs anti-oxytoxic activity and useful in the treatment of ophthgalmia and as a tonic to prevent abortion. It is also reported to be alternative, anti-leprotic, antidropsical, astringent, blood purifier, dentrifice, expectorant, febrifuge, haemostatic, stomachic, styptic and suppurative.

Syzygium cumini (L.) Skeels.

Family : Myrtaceae
Trade/Common Name : Jamun

A tall evergreen tree. Leaves are leathery, smooth, shining to 20 cm long. Flowers are small, dull white, appear in large bunches. Fruits upto 4 cm long, ovoid, black when ripe, juicy and edible.

It is found in throughout moist deciduous forest of India. It is propagated by seeds.

Fresh bark juice mixed with milk is used in diarrhoea. The seed powder about 15 gm

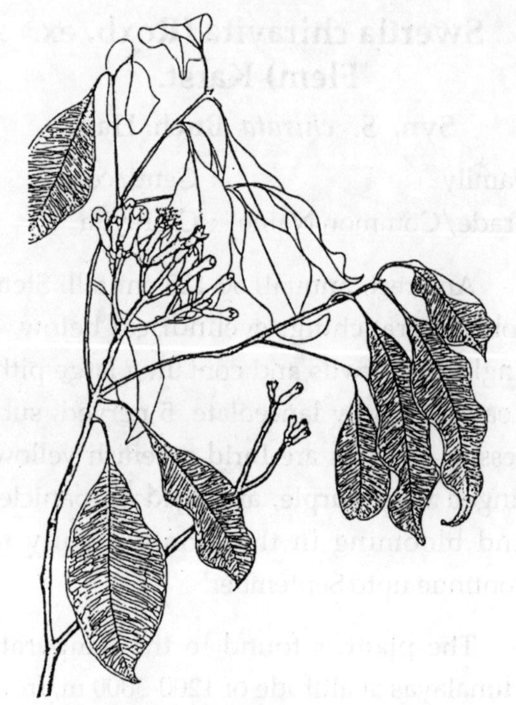

Syzygium cumini

is administered orally thrice a day for 3-4 months in diabetes. The bark is used in sore throats., bronchitis, asthma, ulcers and dysentery.

Tabernaemontana coronaria (Jacq.) Willd.

Syn. *Ervatamia coronaria* Stapf.

Family : Apocyanaceae
Trade/Common Name : Adam's Apple

A very popular shrub with grayish stem and dichotomous branches, grows to a height of 2.5 m. Leaves are membranous, opposite, unequal, simple, elliptic oblong or oblanceolate up to 12 cm long and 6 cm wide; dark green above and dull green beneath. Flowers are odorous at night, white, up to 5 cm in diameter, appear in

Tabernaemontana coronaria

terminal or axillary cymes, blooming in summer and rainy season profusely.

It is cultivated in throughout India. It is propagated by cuttings of ripe shoots or air-layering.

Flowers and wood are fragrant, and essence, used for medicinal purposes.

Tagetes minuta Linn.

Syn. *T. glandulifera* Schrank.

Family : Asteraceae
Trade/Common Name : Jangli-gainda

A highly aromatic annual, 1-2 m tall. Leaves are 7-15 cm long, pinnatisect and the segments are 11-19;4 cm or more long, linear or lanceolate. Flower heads are yellow, narrow, tubular, in corymbose clusters and achenes are black and blooming in the month of July to continue upto September.

The plant is native to South America, is naturalized near waste places and on dry embankments in the north-west Himalaya, between altitudes of 1250-2500 m. It is propagated by seeds.

The flower heads yield an essential oil. It contains aromadendrenem tagetone, phenylethyl alcohol, ocimens, eudesmol etc. It has synergistic activity. It is used as a stomachic, aperient, diuretic and diaphoretic. The ethanolic extracts of the entire herb show anti-viral activity against ranikhet-disease virus.

Tamarindus indica Linn.

Family : Fabaceae
Trade/Common Name : Tamarind

A perennial slow growing tree and more or less evergreen in nature. But leaves fall after the harvest of fruits. Leaflets measure upto 15 cm long and 10-20 in pairs. It has spreading branches and grows to a height of 15 m or even taller. Flowers are appear in clusters and takes 8-10 months to harvest fruit after flowering.

It is native to Africa and grows in India on a commercial scale. It is propagated by seeds or cuttings.

Tamarindus indica

Ripe fruit contains pulp (55%) and it contains tartaric acid, citric acid, malic acid etc. The leaves are applied to reduce inflammatory, swelling and ringworm. Flowers are appetizing, sweet and sour. The seeds are astringent, aphrodisiac and useful in giddiness and vertigo.

Tanacetum dolichophyllum Kitam.

Syn. *T. longifolium* Wallich ex DC.

Family : Asteraceae
Trade/Common Name : Long Leaved Tansy

A perennial, aromatic herb upto 40 cm tall with a ribbed, stout rootstock. Leavesmuch-dissected. Flower heads yellow, larger, rounded forming terminal clusters and blooming in the month of August-September.

It occurs in open slopes, wastelands, forest shrubberies from Pakistan to West Nepal. It is propagated by root cuttings.

It is used as a folk medicine against swelling gums. The water extract of leaf is effective. Also used as an antiseptic.

Tanacetum vulgare Linn.

Family : Asteraceae
Trade/Common Name : Tansy

A perennial herbaceous with strong odour and bitter taste, up to 1.2 m tall

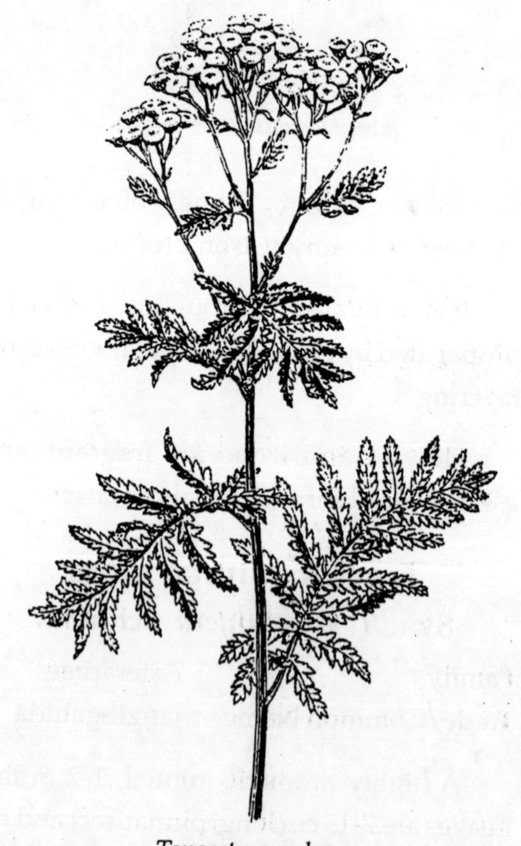

Tanacetum vulgare

grooved, angled, leafy. Leaves are finely divided fan-like up to 12 cm long. Flowers are button-like, yellow in colour, corymbose and pendunde stout.

It is cultivated in Northern parts of India. It is propagated by seeds.

Flowers yield an essential oil (0.2-0.5%), yellowish colour with particular aromatic flavour. It contains beta-thujone, 1-camphor, borneol, thujyl alcohol and terpenes. It is used as anthelmintic, but toxic and overdose causes diarrhoea, vomiting and may be fatal.

Taraxacum officinale Wigg.

Family : Asteraceae
Trade/Common Name : Dudhli

A small herb, 2-10 cm tall, with milky juice. Leaves are all radical, sessile, variable in shape, irregularly pinnatifid and toothed. Heads are 1.0-2.5 cm in diameter, solitary on a hollow and leafless stalk, sessile and bright golden yellow in colour and blooming in the month of March to continue upto November.

The herb is found on grasslands, lawns, roadsides, pavements, vegetable fields, as weed of crop field, throughout the temperate Himalaya, in the Khasi hills of Arunachal Pradesh and in the hills of south India at altitudes of 300-5400 m. It is propagated by seeds.

Roots contain a bitter crystalline principle, taraxacin and a crystalline substance, taraxacerine besides saponin. The leaves contain vitamin C. The root is diuretic, tonic, laxative, cholagogue, anti-rheumatic and asperient. It is a reamdy for chronic disorders of kidney and liver. It is also useful in gallstone, cholecystitis, jaundice, atonic dyspepsia with constipation, muscular rheumatism and oligurea.

Taraxacum officinale

Taxus wallichiana Zuec.

Syn. *T. baccata* Hook.f.

Family : Taxaceae
Trade/Common Name : Himalayan Yew

An evergreen tree, usually 6 m or more

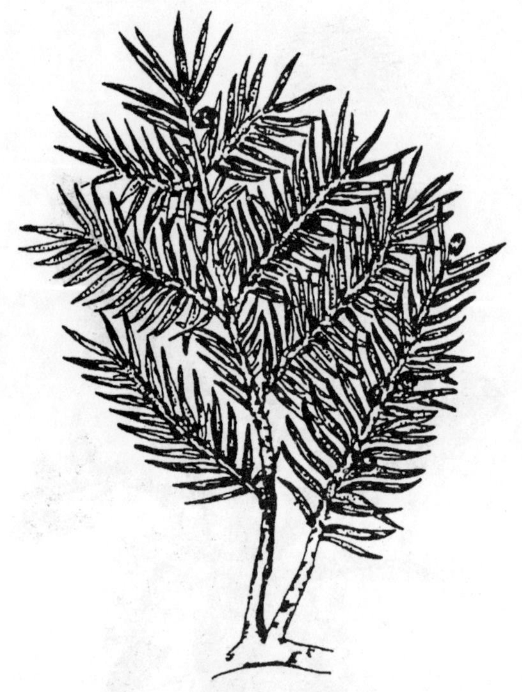

Taxus wallichiana

tall and 1.5-1.8 m in girth. Bark is reddish brown, thin and scaly. Leaves are distichous, linear, with recurved margin, shining above, pale-yellowish brown or rusty red below; Flowers are usually dioecious and the male strobili are stalked, globose, arising from the axils of the leaves on the underside of the branchlets of the previous year; Female strobili are axillary solitary and blooming in the month of March to continue upto May.

The tree is found on forests, shady ravines in the temperate Himalaya at altitudes, between 1800-3300 m and in the hills of Meghalaya and Manipur at an altitude of 1500 m. It is propagated by seeds.

The leaves (needles) contain taxine, taxinine, hydrocyanic acids etc. The young shoots are used for the treatment of headache, giddiness, feeble and falling pulse, coldness of the extremities, diarrhoea and several biliousness. The leaves are employed for the treatment of hysteria, epilepsy and narveness and as a lithic in calculus complaints. Taxol, obtained from the needles is reported to cure uterine and breast cancers.

Tephrosia purpurea (L.) Pers.

Syn. *Cracca purpurea* Linn.

Family : Fabaceae
Trade/Common Name : Sarpankha

A glabrescent, erect undershrub upto 1 m height. Leaf rachis 4-6 cm long,; leaflets 7-17, oblanceolate, obtuse. Stipules narrowly triangular. Flowers bluish-pink to purple, in axillary, leaf opposed, many flowered racemes and blooming in the month of September-October.

Tephrosia purpurea

It is found edges of scrub jungles crop fields, water courses of submontane Himalaya and adjacent plains. It is propagated by seeds.

The plant extract used in hepatic disorders; decoction of pods as vermifuge. Plant as well as roots are considered tonic, stimulant and laxative and used to treat worms and syphilis. A decoction of roots is used to treat chronic coughing, bladder trouble and intestinal worms.

Terminalia arjuna W. & A.

Family : Combretaceae
Trade/Common Name : Arjuna

A large tree with smooth and thick bark, greenish white in colour and flaking off in large flat pieces. Leaves are usually sub-opposite, oblong or elliptic-oblong, pale-brown beneath. Flowers are sessile, in short, axillary spikes or in terminal panocles and blooming in the month of April-May.

It is found throughout the greater parts of India in sub-Himalayan tract, Madhya Pradesh, Chhatisgarh, Bombay and Madras chiefly along water channels stream banks and moist places. It is propagated by seeds.

The bark contains a crystalline compound arjunin, a lactone-arjunetin, essential oil and tannin. It is used as tonic, astringent, febrifuge and deobstruent. It also used in bilious affections and for sores. It is reported as cardiac tonic and useful in blood dysentery, blood pressure, fractures of bone, heart diseases, leucorrhoea and earache, and acts as an antidote to poisons.

Terminalia arjuna

Terminalia bellirica (Gaertn.) Roxb.

Family : Combretaceae
Trade/Common Name : Bahera

A large decidous tree, 20-30 m tall. Leaves are alternate, stalked, broadly ovate with cuneate base, clustered a the end of the branches, and are entire and smooth. Flower are small, pale-green in colour and have offensive odour and blooming in the month of April to continue upto June.

The tree is found in the deciduous forests throughout the greater parts of India,

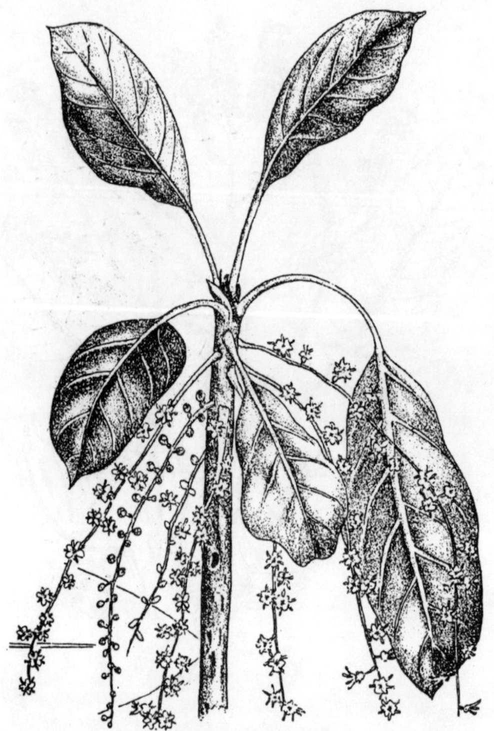

Terminalia bellirica

but not in the arid regions. It is propagated by seeds.

The fruits contain tannins, α-sitosterol, gallic acid, chebulegic acid etc. It is bitter, asterigent, anthelmentic, pungent, laxative and are used in piles, enlargement of spleen, diarrhoea, eye troubles, leprosy and headache.

Terminalia chebula Retz.

Family : Combretaceae

Trade/Common Name : Harad

A large deciduous tree, with thick, dark brown; Young parts covered with rusty hairs. Leaves are mostly sub-opposite, ovate or elliptic, acuminate. Flowers are dull white or yellow, small, strong-scented in terminal, paniculate spikes and blooming in the month of April to continue upto June.

The tree is found in the outer Himalayas ascending upto 1600 m from Himachal Pradesh to Bengal and then throughout the Central and South India. It is propagated by seeds.

The fruits contain tannin (30%), chebulic acid (3.5%), chebulinic acid (30%), ellagic acid, gallic acid and resin. It is alternaive, astringent, dentrifrice, purgative, stomachic and tonic. It is useful in apthae, acsites, asthma, bile trouble, bleeding and ulceration of the gum, blood pressure, burns and scalds, carious teeth, cooling wash of the eyes, cough, dysentery, diarrhoea, piles, vaginal discharge, ulcers, vomiting, and worms.

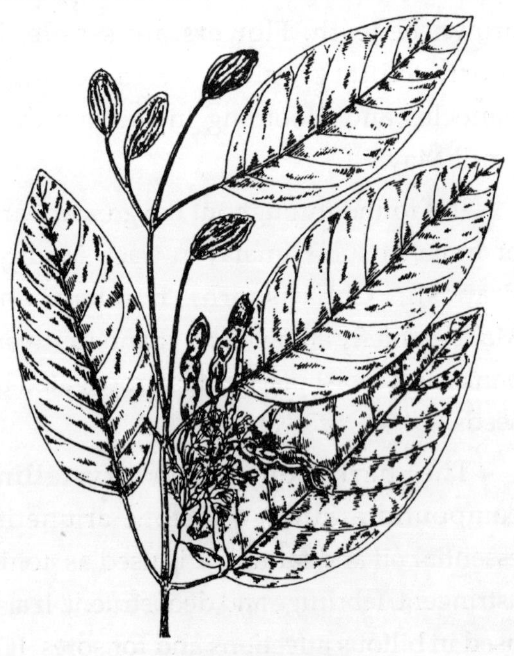

Terminalia chebula

Thalictrum foliolosum D.C.

Family : Ranunculaceae
Trade/Common Name : Mamira

An erect, rigid, perennial herb upto 2.5 m tall. Rootstock is fibrous, yellowish-brown, resem-bling liquorice, but in knotty and extremely bitter and yellow inside. Leaves are pinnately decompound; leaflets 4-6 mm, sub-orbicular, sightly lobulate. Flowers are usually pale-green often dingy-purple, arising in branched terminal panicles and blooming in the month of June to continue upto September.

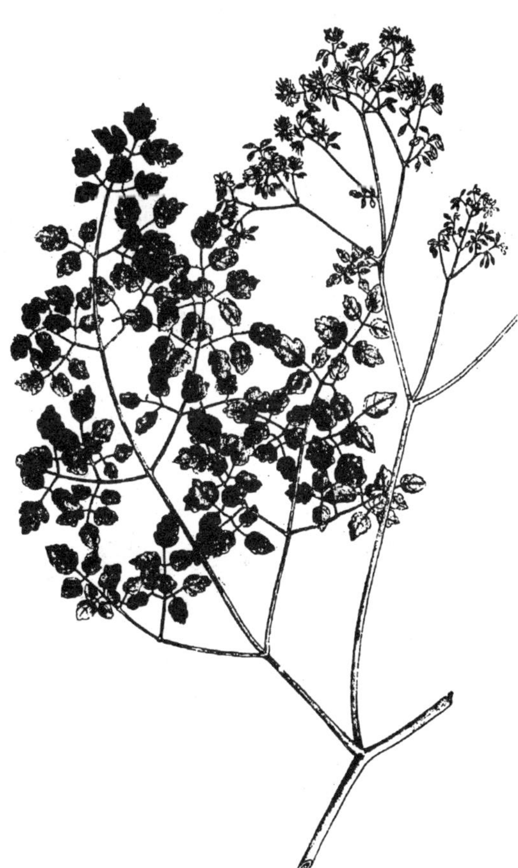

Thalictrum foliolosum

It is found in the inner valleys of the temperate Himalayas and west Tibet between 1800-3500 m. It is propagated by seeds and stem cutting.

The rhizome contains berberine (0.35%), thalictrine, palmitine (0.03%) and used as an antiperiodic, diuretic, aperient and purgative; as bitter tonic. The paste of the plant is locally used in skin diseases and to suppress boils.

Thevetia neriifolia Juss.
Syn. *T. peruviana* (Pers) Merrill.

Family : Apocynaceae
Trade/Common Name : Yellow oleander

A large, evergreen shrub, 4.5-6.0 m tall. Leaves are 10-15 cm long, linear and acute. Flowers are bright yellow in terminal cymes,

Thevetia neriifolia

softly scented and blooming in the month of July-August.

It is found throughout India in field borders, river banks and wastelands. It is propagated by seeds.

The bark contains glucosides thevetin, thevetoxin, a phytosterolin. It is a bitter tonic, cathartic, febrifuge and useful in different kinds of intermediate fevers.

Thymus serpyllum Linn.

Family : Lamiaceae
Trade/Common Name : Wild thyme

An aromatic, hairy, prostate and evergreen perennial herb; 10-25 cm high and is one of the most effective ground covers, forming dense mats. Rootstock is woody and the leaves are small, elliptic or oblong-ovate or lanceolate, 6-15 mm long, subsessile and glandular. Flowers are pink, in globose, terminal whorls and blooming in the month of April to continue upto August.

Thymus serpyllum

The plant is found on sunny hillsides, pasturelands, snady soil, rocks, open woods, clearings in the western Himalayas from Kashmir to Nepal at altitudes of 1500-4500 m. It is propagated by seeds and divisions.

Leaves and floral tops yield a volatile oil (0.27%), scutellarin heteroside, luteolin 7-glucoside and apigenin. It possesses anti-spasmodic, anti-septic, expectorant, carminative, anthelmintic and stimulating properties. The leaves and floral shoots are employed for the the treatment of whooping cough, epilepsy, suppression of urine and menstrual catarrh.

Thymus vulgaris Linn.

Family : Lamiaceae
Trade/Common Name : Common Thyme

A perennial, sub-shrub, attains 45 cm height with hairy branches. Leaves are

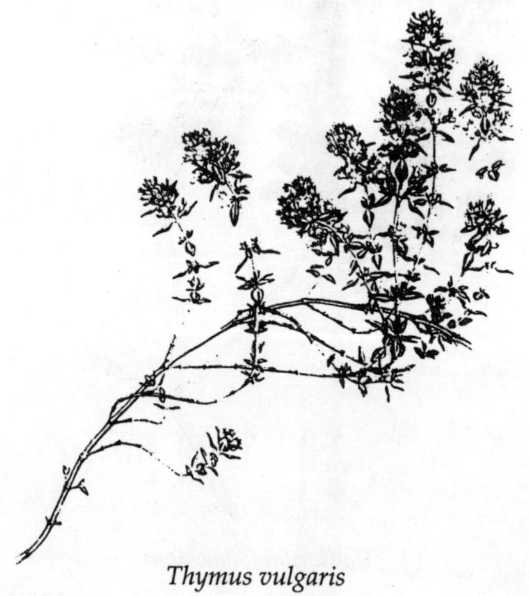

Thymus vulgaris

aromatic and small. Flowers appears in clusters, tubular, lavender and small. Seeds are round and tiny.

It is native to Southern Europe and grown in Northern India upto 900 m altitude. It is propagated by seeds or cuttings.

The decoction of the plant is used as an efficacious in colds, cough, fever and stomach problem. It is used as tonic and spasmodic carminative.

Tinospora cordifolia (Willd.) Miers.

Family : Menispermaceae

Trade/Common Name : Giloe

A glabrous, climbering, succulent shrub with corky grey dotted bark. Leaves are 10-20 cm in diameter, broadly ovate, deeply cordate and shortly acuminate. Flowers are small and greenish-yellow on the old wood in 7.5-1.5 cm long raceme and blooming in the month of March to continue upto June.

The climber is found throughout the tropical regions of India. It is propagated by stem cutting.

Stem contains tinosporin and a furanoid diterpene, giloin, giloinin, berberine and waxy substarte . It is an alternative, anthelmintic, anti-arthritic, antiperiodic, anti-pyretic, aphrodisiac, bitter tonic, blood purifer, cardiac, carminative, digestive, diuretic and expectorant.

Tribulus terrestris Linn.

Family : Zygophyllaceae

Trade/Common Name : Meitcher Kund

An annual prostarte herb upto 70 cm tall. Stem slender, prostrate, branched, nodes swollen slightly purplish, hairy, hair white. Leaves pinnate compound with 4-12 obovate leaflets. Stipules leafy. Flowers solitary axillary in long racemes, petiolate

Tinospora cordifolia

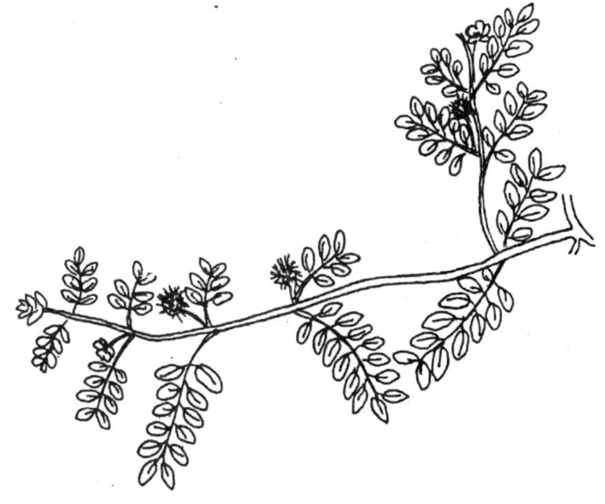

Tribulus terrestris

small, yellow and blooming in the month of July to September.

It occurs in orchard, maize field, vegetable gaedens, kitchen gardens, drylands throughout Warmer regions of the world. It is propagated by seeds.

The herb possesses aperient, tonic, stomachic, aphrodiasic properties. As a folk medicine, leaves are given for the treatment of stones in the bladder. The fruits are usefyl in urinary disorders and sexual weakness. Infusion useful in gout, painful micturition.

Trigonella foenum-graceum Linn.

Family : Fabaceae
Trade/Common Name : Ferugreek

A hardy annual plant. Leaves are feather shaped. Flowers are whitish, in colour.

Trigonella foenum-graceum

It is native to Eastern Europe. It occurs wild in parts of North India. It is propagated by seeds.

Seeds are used as condiment and dried leaves for flavouring. The vegetative parts are rich in vitamin A, B, and iron. The mucilaginous seeds are used in medicine. The seeds are also used to resolve inflammatory tumours, such as carbuncles, phlegmous and whitlows. The flower is recommended for aphtha, chapping, haemorrhoidal tumours, in rectal injection, for dysentery, diarrhoea, bilious and inflammatory colics.

Trillidium govanianum (D. Don.) Kunth.

Syn. *Trillium govanianum* Wall. ex D. Don.

Family : Trilliaceae
Trade/Common Name :

A tuberous rooted herb often upto 30 cm tall with three leaves in one whorl at the summit of the stem and a solitary, purple flower in the center. Leaves are broadly ovate, acute and conspicuously stalked. Flowers brown purple with narrow and blooming in the month of May-June.

It is found in patches in forest shade rich in humus from Pakistan to Bhutan between 2700-4000 m in forests. It is propagated by seeds.

Roots contain trillarin, which on hydrolysis yield 2.5% diosgenin-a

Trillidium govanianum

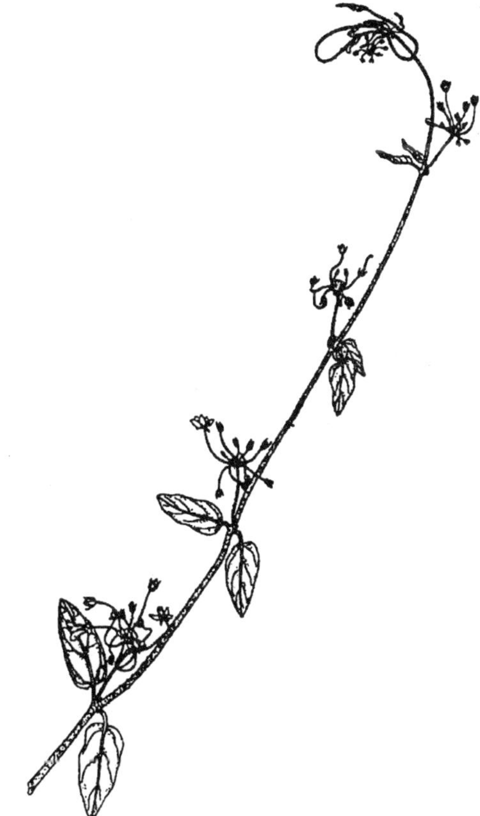

Tylophora indica(Burm. f) Merr.

corticosteroid hormone. It is used in various preparations like sex hormones, cortisone and allied preparation used in rheumatism, birth control, regulation of menstrual flow etc.

Tylophora indica (Burm. f.) Merr.

Syn. *T. asthmatica* W. & A.

Family : Asclepiadaceae
Trade/Common Name : Antamul

A small, evergreen, climbing shrub with pubescent branchlets. Leaves opposite, ovate or ovate-lanceolate. Flowers small, few flowered, solitary or clustered umbels.

It is found dry forest of sub-Himayalan tracts of North and East Bengal, Assam, Cachar, Orissa, Konkan, Deccan, and plains of Tamil Nadu. It is propagated by seeds or shoot cuttings.

Dry plant yields 0.44% total alkaloids containing tylophorine (0.1%). It is used as emetic, diaphor, expectorantused in bronchial, cold and cough. Roots regarded as abortifacient. It is used as substitute for Ipecacuanha.

Ulmus wallichiana Planchon.

Family : Ulmaceae
Trade/Common Name : Mairu

A deciduous tree upto 30 m tall. Bark rough, grey, exfoliating in diamond-shaped scales. Branchlets tomentose. Leaves alternate, obliquely elliptic-ovate or obbovate. Flowers red-brown, in dense lateral racemes appearing before the leaves and blooming in the month of March-April.

It is found in mixed deciduous forests of montane zones of Kashmir to Nepal. It is propagated by seeds or division.

Plaster used for fractured bones is made by boiling the bark in water.

Ulmus wallichiana

Uncaria scandens (Smith) Hatch.

Syn. *Nauclea scandens* Smith; *U. pilosa* Roxb.

Family : Naucleaceae
Trade/Common Name : Gambier

An evergreen, scandent shrub, climbing by means of recurved hooks. Leaves opposite, ovate or elliptic, abruptly cordate-acuminate, rounded at base. Flowers in globose heads, on axillary stout peduncles, with a whorl of linear, lanceolate bracts above the middle and blooming in the month of June-August.

It is found in Sub Himalayan tracts, nearly ravines from Garhwal to Sikkim. It is propagated by seeds or division.

Leaves contains catechins (7-20%). It posses astringent properties and used to treat diarrhoea and sore throat.

Uraria picta (Jacq.) Desv. ex DC.

Family : Papilionaceae
Trade/Common Name : Chitraparni

An erect, little branched, perennial herb, 90-180 cm tall. Leaves are very variable, 1-3 foliate, upto 30 cm long and blotched white. Leaflets are 4-6, rarely 9, linear, oblong or lanceolate, obtuse and mucronate with subulate stipules. Flower are purple in colour, dense, long, cyclindrical, terminal raceme and blooming in the month of July-September.

Uraria picta

drooping or spreading distant in a terminal raceme and blooming in the month of May to continue upto September.

It is found in Western Himalaya below 1400 m; Bihar, Jharkhand, Konkan and Coromandal. It is propagated by seeds.

The fresh bulbs yield two glucosides, scillarin-A, which is crystalline and scillarin-B, which is amorphous. Its action is anthelmintic, cardiac, stimulant, deobtruent, digestive, diuretic, emmenagogue and expectorant and useful in asthma, ailments if hearts, calculous affections, cough and

The plant is found in the dry grasslands, waste places and open forest in the sub-Himalayan tract from Kashmir ot West Bengal and Assam, upto an altitude of 1800 m, and all over the plains of India. It is propagated by seeds.

The plant is one of the components of 'Desmoola' and is credited with fracture healing properties. The root possesses aphrodisiac properties and its decoction is prescribed for cough, chills and fevers.

Urginea indica (Roxb.) Kunth.

Syn. *Scilla indica* Roxb.

Family : Liliaceae
Trade/Common Name : Jangli-piyaz

A glabrous herb. Bulb dull white or pale, ovoid. Leaves radicle, flat, narrow pointed. Flowers appearing before the leaves,

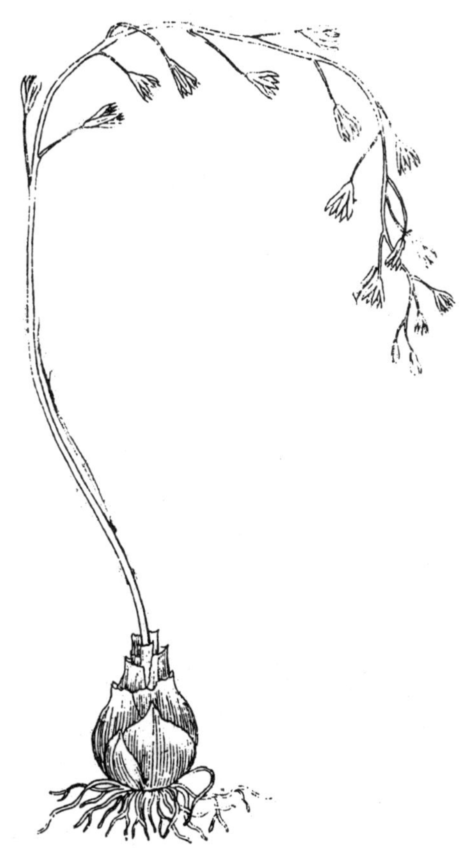

Urginea indica

bronchitis, paralytic affections, promotes urination, in rheumatism and skin diseases. It also employed in dropsy, rheumatism and skin troubles.

Urtica dioica Linn.

Family : Urticaceae
Trade/Common Name : Stinging Nettle

A biennial or perennial herb upto 1 m tall. Stem branched from the base, angular, grooved, ribbed all over covered with stinging hairs. Leaves simple, ovate to obovate, covered with stinging hairs. Flowers unisexual, green, clustered in long or short thread-like catkins and blooming in the month of August-October.

Urtica dioica

It occurs in wastelands, fallow places, margins of cultivated fields, woodlands, shrubberies in Western Himalaya. It is propagated by seeds.

Water extract of leaves has antiseptic properties. Used to make shave lotions and shampoos.

Uvaria odorata Linn.

Family : Anonaceae
Trade/Common Name : Pulichan

A shrubby climber or straggling bush with stout stem. Leaves are ovate-lanceolate, up to 12 cm long and 4 cm broad, alternate and upper surface shining. Flowers are pendulous, solitary, yellowish green in colour and fragrant.

It is native of India. It is propagated by layering.

Flowers are used for perfumes.

Vaccinium myrtillus Linn.

Family : Ericaceae
Trade/Common Name : Bilberry

A hardy deciduous shrub attains a height upto 45 cm. Flowers are pale pink in colour. Fruit is berry and acidic in nature.

It is native to Britian. It is common in the hilly districts of India. It is propagated by seeds or division.

Fruit are sedative and astringent and its preparation are used in the treatment for diarrhoea and dysentery.

Vaccinium myrtillus

Valeriana hardwickii Wall.

Family : Valerianaceae
Trade/Common Name : Muskbala

An erect, pubescent herbs, 20-60 cm tall. Radical leaves few, ovate-orbiculate, entire or crenulate, pilose, cauline leaves 3-7 pinnate, terminal lobe the largest. Lobes ovate-lanceolate, creanate, serrate. Flowers white in axillary or terminal paniculate cymes or corymbs and blooming in the month of June to continue upto August.

The plant is found from Pakistan to South-west China, Burma, South-east Asia, frequent on moist slopes in the Himalayas and Khasi hills in Shrubberies and open

Valeriana hardwickii

slopes between 1500-4000 m elevation. It is propagated by seeds and multiplying rootstock.

Roots contain a crystalline acid, α-eudesmol, α-eudesmol, edemol, angelicin etc. It is used as insecticide and useful in mental disorder, scenting hair.

Valeriana jatamansi Jones.

Syn. *V. wallichii* DC.

Family : Valerianaceae
Trade/Common Name : Indian Valerian

An erect, pubescent herb, having horizontal, thick rootstock with thick, descending fibrous roots; stem 15-45 cm high, usually tufted. Basal leaves are deeply cordate, ovate, usually acute and toothed. Stem leaves are much smaller, few, entire or pinnate. Flowers are white or tinged with

Valeriana jatamansi

pink in terminal corymbs and blooming in the month of March to continue upto May.

The plant occurs abundantly on open slopes, grasslands, forest and shrubberies in the temperate Himalayas from Kashmir to Bhutan at altitude of 1300-3800 m and in the Khasi hills between 1250-1800 m. It is propagated by seeds.

Indian valerian yields a volatile oil (0.5-2.2.12%). It possesses aphrodisiac, antiseptic, cardiac, stimulant, carminative, diuretic, emmengogue, expectorant, febrifuge, nervine tonic, ophthalmic, sedative, suppurative and tonic properties. It is useful in hysteria, epilepsy, chorea, shell shock and neurosis.

Valeriana officinalis Linn.

Family : Valerianaceae
Trade/Common Name : Common Valerin

A perennial glabrous herb, attain a height upto 1 m. Stem solitary, erect, furrowed and hollow. Leaves are opposite, pinate, lower leaves with long petioles and

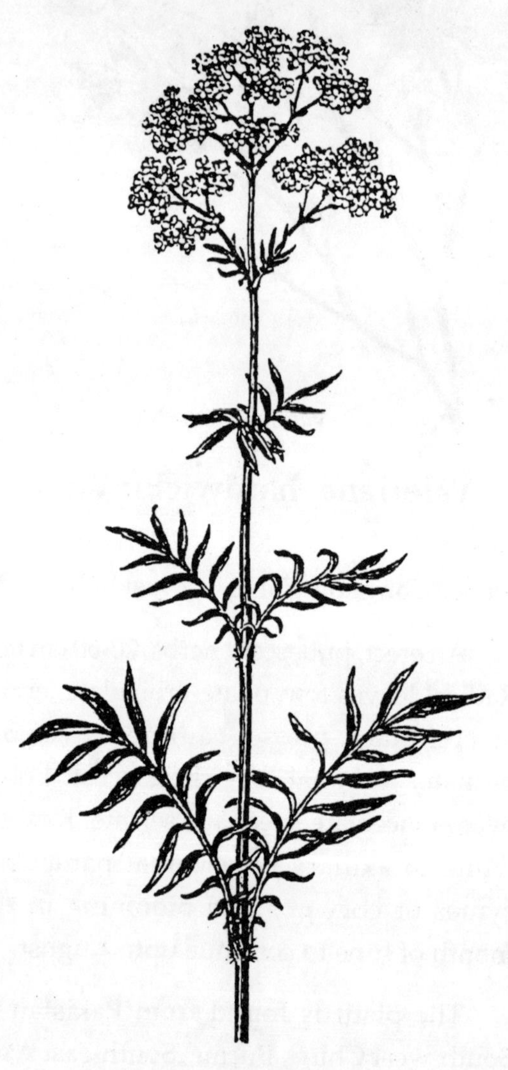

Valeriana officinalis

upper small. Flowers appear in Cymose panicle, corymb, white or dull white in colour, fragrant, small, smooth, without hairs, on seeded.

It is native to Europe. It is found in Kashmir at 2500 m altitude. It is propagated by seeds or through rootstock.

Root contains epoxy-iridoid esters called valepotriates which include valtrate didrovaltrate, acevaltrate, isovaltrate etc. The drug is said to be sedative, hypotensine, antispasmodic and stomachic. The juice of fresh rhizomes is used in treatment of hysterical fits and other nervous disorders and flatulence. The root preparation is also known as tonic and stimulant. The essential oil of the root is used as a flavour in food product and beverages.

Vanda tessellata (Roxb.) Hk. ex D. Don

Syn. *V. roxburghii* R. Br.

Family : Orchidaceae
Trade/Common Name : Vanda

An epiphytic orchid, 20-30 cm high, with leafy stem. Leaves are thickly cariaceous, recurved, plicate, obtusely keeled. Flowers are greenish yellow, mottled with brown on the mid lobe of lib with purple caruncles and blooming in the month of April to continue upto June.

It is found in Bengal, Jharkhand, Bihar, Madhya Pradesh, Chhatisgarh, West Peninsula and Travancore. It is propagated by cutting.

The plant contains an alkaloid, glucoside, bitter principle, tannins, resin, saponin, sitosterols etc. In action, It is anti-arthritic and useful in rheumatism and allied disorders. Paste of the leaves is applied to the body of fever. Leaf juice is a remedy for otitis media.

Verbascum thapsus Linn.

Family : Scrophulariaceae
Trade/Common Name : Cow's Lungwort

A annual herb, leafy, unbranched stem upto 1.5 m in height. Basal leavesstalked, simple, clothed with a slender woolly tomentum all over. Upper leaves smaller, clasping the stem. Flowers are yellow

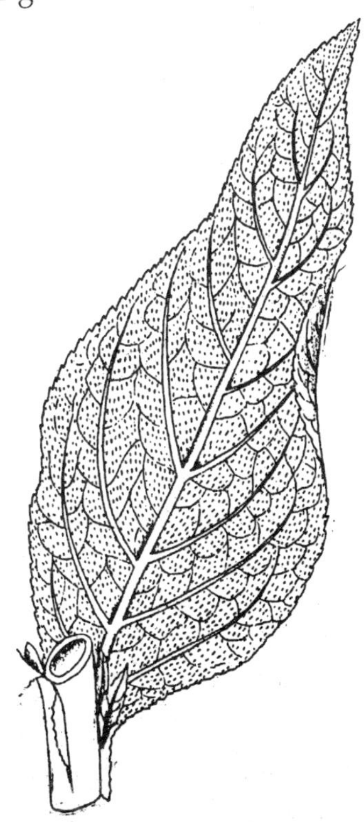

Verbascum thapsus

arranged on long apical spike and blooming in the month of January to June.

It is found in stony slpoes, wastelands, fallow places, dry situation in temperate Himalayas from Kashmir to Bhutan. It is propagated by seeds.

Flowers contain corcetin. Leaves contain saponins. Roots used as febrifuge. Its decoction is administered in cramps and migraine. The herb is useful for the treatment of asthma and other pulmonary complaints. Seeds are aphrodisiac and narcotic in nature.

Verbena officinalis Linn.

Family : Verbenaceae
Trade/Common Name : Vervain

An annual herb upto 60 cm in height. Stem having ascending branches, downy or puberulous. Leaves simple to pinnatifid. Flowers in long apical spikes, white or slightly pinkish and blooming throughout the year.

It occurs in streamsides, orchards, wastelands, vegetable fields, forest clearings in temperate and sun-tropical regions of Himalaya from Kashmir to Bhutan. It is propagated by seeds.

It contains verbenalins. Leaves are febrifuge and tonic in nature and used in rheumatism and diseases of joints. Its ointment is given for swelling of womb. Root is remedy for scrofula and snakebite.

Verbena officinalis

Veronica officinalis Linn.

Family : Scrophulariaceae
Trade/Common Name : Speedwell

A hardy perennial herb.

It is native to Europe and Asia. It is propagated by seeds or division.

Herb is used as light tonic in the convalescence of certain fevers, asthma and bronchitis. It is act as a stimulant, provoking urination and facilitating expectoration. It is also used for the treatment of chest

Podophyllum hexandrum

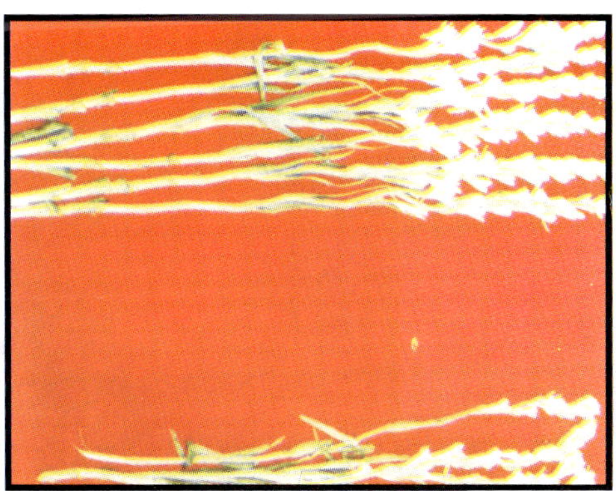
Polianthes tuberosa

Polygonum alpinum

Polygonum hydropiper

Portulaca oleracea

Prangos pabularia

Prunella vulgaris

Punica granatum

Rhododendron arboreum

Rauvolfia serpentia

Ricinus communis

Rosa damascena

Rosa sericea

Rosmarinus officinalis

Santalum album

Saraca indica

Saussurea gossypiphora

Saussurea obvallata

Senecio jacquemontianus

Solanum nigrum

Solanum surattense

Spilanthes acmella

Stellaria media

Swertia chirata

Syzygium cumini

Tamarindus indica

Tanacetum dolichophyllum

Taxus baccata

Tephrosia purpurea

Terminalia arjuna

Termamalia ballarica

Terminalia chebula

Thevetia neriifolia

Tinospora cordifolia

Tribulus terrestris

Trigonella foenum-graecum

Urginea indica

Valeriana jatamansi

Verbascum thapsus

Verbena officinalis

Vitex negundo

Withania somnifera

Xanthium strumarium

Zingiber officinale

Veronica officinalis

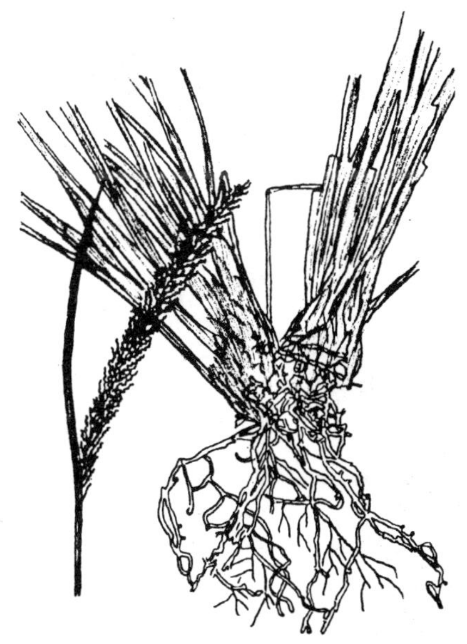

Vetiveria zizanioides

diseases, such as chronic pulmonary catarrh, phthisis and bronchitis.

Vetiveria zizanioides (Linn.) Nash.

Family : Poaceae
Trade/Common Name : Vetiver/Khas

A densely tufted awnless, wiry, glabrous, perennial grass; is a shy breeder and is considered sterile. A large clumps from a much branched spongy rootstock with erect clums, 0.5-1.5 cm high. The leaf blades are relatively stiff, long and narrow upto 75 cm long and 8 mm in width, glabrous, but, downward rough along the edges. Flowers in conical panicles, 15-30 cm long, narrow, acute, appressed, awnless, sessile and somewhat flattened laterally with short, sharp spines and blooming in the month of July to continue upto September.

The grass is found throughout the plains in lower hills of India, particularly on the riverbanks and in rich marshy soils, ascending to an altitude of 1200 m. It is propagated by slips.

Roots yield an essential oil (0.2-1.0%). It is aromatic, refrigerent, febrifuge, diaphoretic, stimulant, stomachic, emmenagogue, expectorant, heart-tonic, nerve sedative and diuretic. Root part is cooling and essential oil acts as tonic and useful to allay thirst in fever and inflammatory affections.

Viola biflora Linn.

Family : Violaceae
Trade/Common Name : Yellow Violet

An annual, glabrous herb, with sender rootstock; stem decumbent. Basal leaves reniform to orbicular, margin crenate.

Viola biflora

Flowers yellow, solitary or paired and blooming in the month of May-July.

It is found in marshy slopes at Chandrabadni, Bhilagana valley and Ketharnath of Garhwal hills, upto 3,500 m altitude. It is propagated by seeds or rootstocks.

Leaves are used as laxative and the roots as emetic.

Viola odorata Linn.

Family : Violaceae
Trade/Common Name : Sweet violet

A perennial herb with stout creeping rootstock. Leaves all radical, upto 3 cm diam., petiolate, stipules elongated, leafy. Flowers pinkish-blue or purlish with long pedicle and blooming in the month of March to continue upto August.

It occurs in shrubberies, forest clearings, rocky and shady areas, mud wall, hedgebanks in North-West Himalaya. It is propagted by rot cutting.

Rhizome contains glycoside-methyl salicylate, an alkaloid violine, a glycoside-violequarcitin. It posses diuretic and emetic properties and internally used as infusion or as a component of herbal tea for relieving cough and bronchitis. The flowers emollient,

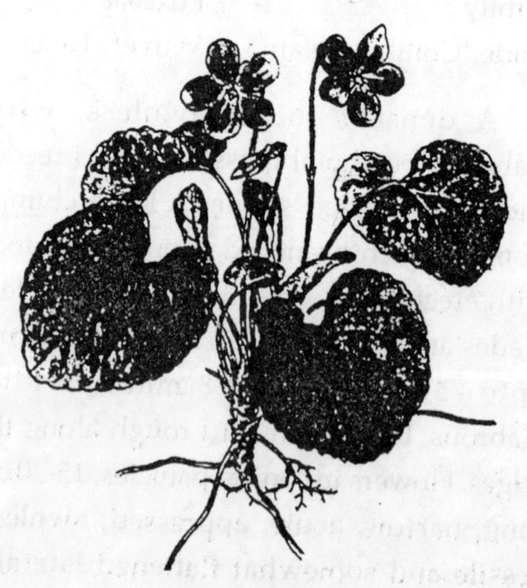
Viola odorata

demulcent, used in biliouness and lung troubles. It also used as blood purifier.

Viola serpens Wall.

Syn. *V. pilosa* Wall. ex. Ging.

Family : Violaceae
Trade/Common Name : Banfshah

A small perennial herb, with short tufted rootstocks. Stem short, but distinct, covered with withered scales, often producing runners. Leaves are ovate-cordate, crenate-serrate. Flowers lilac, blue or sometimes white and blooming in the month of March to continue upto August.

A hirsute or glabrous herb, found throughout the temperate Himalayas upto an altitude of 2000 m. It is propagated by divisions, cutting and seeds.

Viola serpens

Roots contain glucoside methyle salicylate; yield an alkaloid, violine; a glucoside, violquercitrin, saponin etc. It possesses aperient, antipyretic, cooling, demulcent, diaphoretic, diuretic, emetic, emollient, expectorant, febrifuge and purgative. It is useful in asthma, bleeding piles, cancer of the throat, constipation, cough, fever, headache, and skin diseases.

Viscum album Linn.

Syn. *V. costatum* Gamble.

Family : Loranthaceae
Trade/Common Name : Mistletoe

A large, epiphytic, profusely branching shrubby perennial, forming rouned clumps over the deciduous trees. It is green all over; branches dichotomous or whorled. Leaves simple about 5-10 cm, sessile, cuneate-oblong, or oblanceolate, with 3-5 longitudinal, basal nerves, wavy margin, opposite at nodes. Flowers are diocious, sessile; in clustered of 3-5, supported by concave bracts and blooming in the month of May to continue upto September.

Due to parasitic in nature, It occurs on Rosaceous trees, shrubs, on elm, walnut and willows in Uttaranchal and Himachal Pradesh upto 2000 m. It is propagated by seeds.

The herb contains fatty oil (oleic & palmitic), nicotine alkaloids, amines, flavonoids and chaleone derivatives.

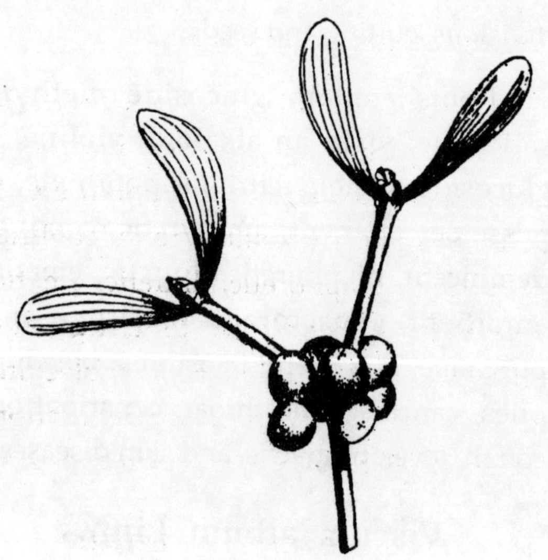

Viscum album

"Viscin", fatty oil, lutein, alpha and beta carotene and vitamin C. It is used for hypertension, arteriosclerosis, tumors, epilepsy, hysteria, ear diseases and arthrosis. It possesses cardiotonic, antiseptic, antispasmodic, diuretic, purgative and emetic with narcotic action.

Vitex negundo Linn.

Family : Verbenaceae
Trade/Common Name : Nirgandi

A shrub, with a thin, grey bark. Branchlets are 4-angled, densely white tomentose. Leaves 3-4 foliate, pediolate. Leaflets lanceolate, coarsely cut & crenate; terminal leaflets quite long; lateral smaller and with much shorter petiolules. Flowers bluishes-white to blue in terminal, paniculate racemes and blooming in the month of March-April.

The plant is found throughout the greater part of India, ascending to an

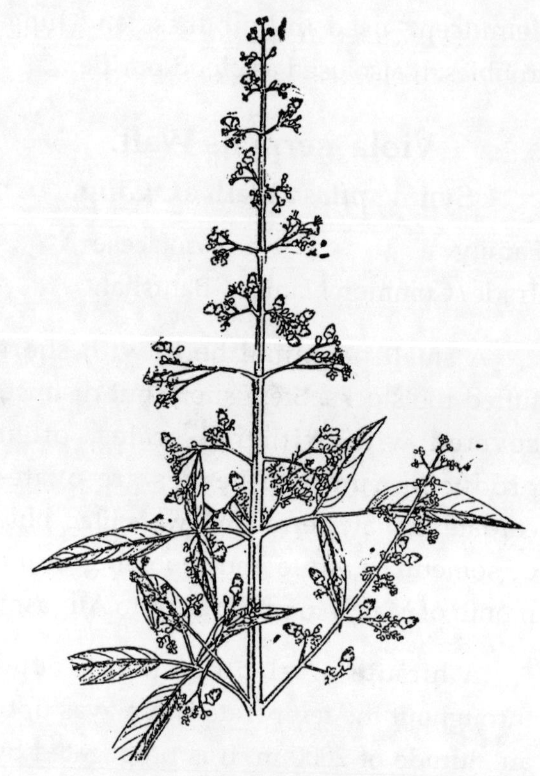

Vitex negundo

altitude of 1500 m in the outer Himalayas. It is propagated by seeds.

Fresh leaves yield an essential oil (0.05%); bark yield sflavone glycosides. The leaves possesses alternaive, anodyne, antiarthritic, antiparasitic, appetizer, aromatic, astrigent, discutient, febrifuge and nerine tonic properties. It is also useful in curative effects in asthma, lung diseases, spenic enlagement, urinary troubles, tonsillitis, sciatica rheumatism etc.

Wisteria sinensis (Sams.) Sweet.

Family : Fabaceae
Trade/Common Name : Kidney-bean Tree

A very free flowering and most popular,

vigorous climber, may climb to a height of 12 m and produce deep lilac spikes. Leaflets are ovate, acuminate, in opposite, distinct pairs, clothed on both sides with thin, silky, adpressed pubescence hairs.

It is native of China. It is found in various parts of India. It is propagated by seeds or cuttings.

Flowers yield concrete (0.2%) by solvent extraction. The concrete is dark brown in colour. It is used for perfume manufacturing.

Withania somnifera Dunal.

Family : Solanaceae
Trade/Common Name : Ashvagandha

An erect, braching perennial undershrub upto 1.5 m in high. All parts are nearly, more or less stellately tomentose and the branches are flexuose, terete and densely tomentose. Leaves are petiolate, 5-10 cm long, ovate, sub-acute. Flowers are greenish or lurid yellow, usually about 5 together in sub-sessile umbelliform cymes and blooming in the month of June to continue upto August.

The plant is found in the dried parts of India, ascending upto 1700 m in Himayalas. It is propagated by seeds.

Roots contain an essential oil ipuranol, a crystalline alcohol, whithaniol, withaninie, somniferene, hentriacontane, phytosterol, etc. It is used as alternative, aphrodisiac, tonic, deobstruent, diuretic, narcotic, hypnotic, sedative, restoration and abortifacient and useful in rheumatism, cough, debility from old age, dropsy, emaciation of children, consuption and general weakness.

Withania somnifera .

Woodfordia fruticosa (L.) Kurtz.

Syn. *W. floribunda* Salisb.

Family : Lythraceae
Trade/Common Name : Dhai-ki-phool

A large, much branched deciduous shrub with bark scaling off in small patches.

Woodfordia fruticosa

Leaves are opposite or in whorls of 3, ovate lanceolate. Flowers and their calyces are bright red, the later permanent, and retain their colour even after the flowers have faded. Flowers are arranged in short, clustered panicles in appear before the leaves and blooming in the month of March to continue upto June.

It is found in dry slopes, rocky fields and waste lands in throughout the greater parts of India upto 1600 m in the Himalaya. It is propagated by seeds.

Flowers yield tannin upto 20%. It possesses asterigent, stimulant, suppurative, styptic properties. It is useful in bowel camplaints, haemorrhages, diarrhoea and dysentery, complaints of the liver, leucorrhoea, menorrhegia, stimulant in pregnancy, skin diseases, vascular eruptions and foul ulcers.

Wrightia arborea (Den) Mabb.

Syn. *W. tomentosa* (Roxb.) Rhem. & Schult.

Family : Anonaceae

Trade/Common Name : Dudhi

A deciduous shrub, with grey corky bark; apical part pubescent. Leaves opposite, elliptic, shortly caudate-acuminate. Flowers are pale-yellow to white, in erect, terminal, corymbose cymes and blooming in the month of May-June.

It is found in miscellaneous forests, often associated with Sal, in sub Himalayan tracts, ascending to 1200 m altitude. It also occurs in Bihar, Assam, Punjab, Rajasthan and Western peninsula. It is propagated by seeds or cuttings.

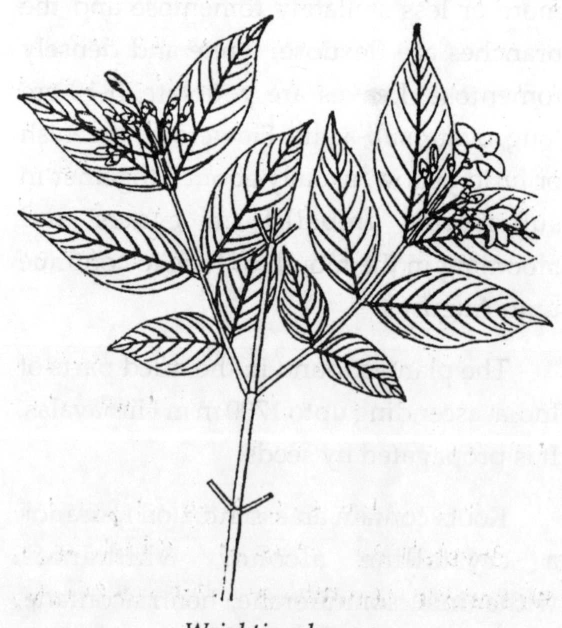

Wrightia arborea

Herb is used for relief from teethache and fever. The milky juice is used to stop haemorrhage. The leaves and tender flowers are edible. Stem and root bark regarded as an antidote to snake-bite and sting of scorpions.

Xanthium strumarium Linn.

Family : Asteraceae
Trade/Common Name : Adhasis/ Burweed

An erect, tall, coarse, unarmed weed with triangular, cordate or sub-orbicular, often 3-lobed, irregularly toothed, scabrid or hispid leaves. Flowers numerous, white or green in terminal and axillary racemes and blooming in the month of April to continue upto September.

It is found throughout India. Widely distributed throughout the world in tropical and temperate region. It is propagated by seeds.

The aerial parts of the plant contain a mixture of alkaloids i.e. sesquiterpene lactones-xanthatin. It is powerful diaphoretic in nature and useful in chronic malaria, leucorrhoea and urinary diseases. The root is bitter tonic, useful in cancer and scrofula and herpes in power form. The fruit is used as tonic, diuretic, diaphoretic, sedative, cooling and demulcent and also given in small-pox.

Xanthium strumarium

Zanthoxylum aromatum DC.

Syn. *Z. alatum* Roxb.

Family : Rutaceae
Trade/Common Name : Tumbru

An armed, scandent or erect shrub upto 6 m tall with dense foliage. Branches-armed and the prickles are flattened, upto 2 cm long. Bark is pale brown and deeply furrowed. Leaves are imparipinnate or trifoliolate. 5-20 cm long, often with lanceolate, entire to glandular, crenate, acute to obtusely acuminate leaflets. Flowers are greenish-yellow in dense, terminal and axillary, sparse panicles and blooming in the month of April-May.

Zanthoxylum aromatum

It occurs in slopes, nalahs, field borders and depressions below 2200 m elevation in Punjab, Uttaranchal, Khasi hills and Himachal Pradesh. It is propagated by seeds.

Dried fruit yields an essential oil (2.3%). The essential oil contains linalool, linalylacetate, citral, geraniol, etc. The seeds and barks are used as an aromatic tonic in fever, dyspepsia and cholera. Fruits are used as a remedy for toothache and carminative, stomachic and anthelmintic in action. The essential oil possess antiseptic, disinfectant, deodorant properties. The twigs of branches are chewed and used as tooth sticks for cleaning teeth. Locally, the seeds are used as a condiment and for water purification.

Root bark is used as flavouring agent in place of mulhatti in pan (betel leaf).

Zanthoxylum clava-herculis Linn.

Family : Rutaceae
Trade/Common Name : Princkly Yellow Wood

A prickly shrub of 2 m tall. Leaves are compound with 5 to 11 leaflets, arranged alternately. Flowers are small, inconspicuous and greenish white in colour.

It is found various parts of India. It is propagated by seeds or cuttings.

Berries, bark, wood and roots are medicinally useful. To treat various kinds of itches, a lotion made by boiling inner bark in water is applied. Berries are considered stimulant, tonic, antirheumatic and also treat colic, gas, muscle spasms and chest ailments. A decoction of bark is useful for treating gonorrhoea.

Zingiber officinale Rose.

Family : Zingiberaceae
Trade/Common Name : Ginger

A herb with tuberous, horizontal and aromatic root stocks. Stem are leafy and 1.8 m tall. Leaves are 30 cm long and 7.5 cm broad. Flowers are white or yellowish in colour, with pale-yellow lip.

It is cultivated in various parts of India. It is propagated by root cuttings.

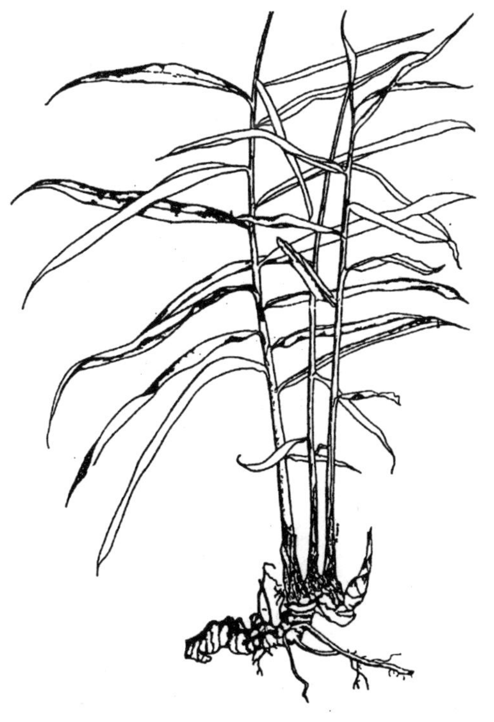

Zingiber officinale

Zizyphus mauritiana

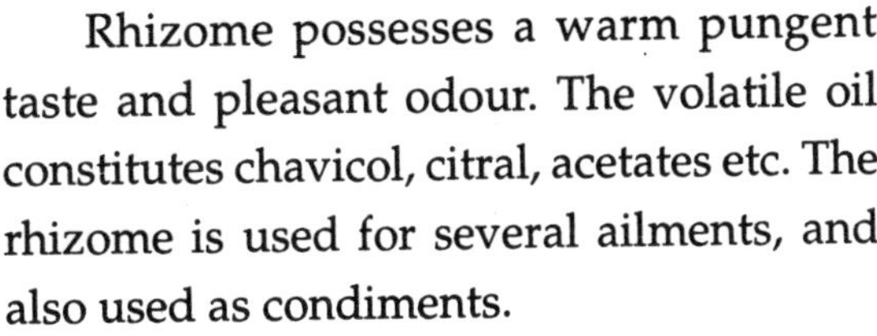

Rhizome possesses a warm pungent taste and pleasant odour. The volatile oil constitutes chavicol, citral, acetates etc. The rhizome is used for several ailments, and also used as condiments.

Zizyphus mauritiana Lam.

Syn. *Z. jujuba* (L.) Gaert.; *Rhamnus jujuba* Linn.

Family : Rhamnaceae
Trade/Common Name : Ber

A subdeciduous , spinous tree or shrub. Stem grey brown, young branches densely tomentose. Leaves varible, elliptic, ovate-oblong, serrulate, base subcordate, ovate-obtuse. Princkles solitary, straight or curved. Flowers in short axillary, sussessile cymes, greenish-yellow and blooming in the month of June-August.

It is commonly distributed in different parts of India and warmer parts of sub Himalayan tracts (Garhwal hills). It is propagated by seeds or cuttings.

Fruit contains vitamin A, B and C, sugar, protein and minerals. In sore throat, a leaf decoction is used as gargle. For dysentery and diarrhoea, decoction of root and bark is prescribed. To treat bleeding gums, leaf decoction is also useful. The seed kernels are considered aphrodisiac.

List of Endangered Medicinal Plants

Sl.No.	Botanical Name	Family	English Name	Part Used	Habit & Habitat	IUCN Status	Active Constituents	Threats	Current Regional Distribution
1.	*Aconitum balfourii* Stapf.	Ranunculaceae	—	Roots	Herb/Sub-alpine to Alpine Himalayas	CR/N	Pseudoaconitine (0.4-0.5%)	1, 2, 3, 4 & 5	Uttaranchal to Nepal Himalayas (3,000 4,500 m altitude)
2.	*Aconitum deinorrhizum* Stapf.	Ranunculaceae	Safed Bish	Roots	Herb/Alpine grassy slopes	CR	Pseudoaconitine (0.51%)	2, 4, 1 & 5	J&K to H.P. Himalayas (3,000-3,500 m altitude)
3.	*Aconitum falconeri* Stapf.	Ranunculaceae	Meetha Tellia	Roots	Herb/Sub-Alpine to Alpine meadows	CR	Pseudoaconitine (0.47%)	1, 2, 3 & 5	Garhwal Himalayas of Uttranchal (3,300-4,000 m altitude)
4.	*Aconitum ferox* Wallich ex Ser.	Ranunculaceae	Indian Aconite	Roots	Herb/Moist forest Temperate to Sub-alpine Himalayas	CR/R	Total alkaloids (0.63-4.7%)	2, 4, 1 & 5	Western Indian Himalayas (2,100-3,800 m altitude)
5.	*Aconitum heterophyllum* Wall. ex Royle	Ranunculaceae	Atis roots	Roots	Herb/Sub-alpine to Alpine regions in grassy meadows	CR/N	Atisine (0.4%), Aconitic acid	2, 4, 3, 1 & 5	J & K to Garhwal regions of Uttaranchal Himalayas (3,000-4,200 m altitude)
6.	*Aconitum violaceum* Jacq. ex Stapf.	Ranunculaceae	Patis	Roots	Herb/Alpine Himalayas	CR/N	Pseudoaconitine (1%)	1, 2, 3, 4 & 5	Indian Northwestern Himalayas (3,000-5,100 m altitude)
7.	*Acorus calamus* Linn.	Araceae	Sweet Flag	Rhizome	Herb/Swampy & marshy open places	VU/R	Essential oil (1.8%)	2, 1 & 5	W. Bengal, Sikkim & Meghalaya (2,000-3,000 m altitude)

Contd...

Table Contd...

Sl.No.	*Botanical Name*	*Family*	*English Name*	*Part Used*	*Habit & Habitat*	*IUCN Status*	*Active Constituents*	*Threats*	*Current Regional Distribution*
8.	*Angelica glauca* Edgew.	Apiaceae	Chora	Roots	Herb/Shady, moist slopes of temperate Himalayas	CR/R	Essential oil (0.4-1.3%)	1, 4 & 5	J & K to Kumaon region of Uttaranchal Himalayas (2,500-3,500 m altitude)
9.	*Aquilaria malaccensis* Lam. Syn. *A. agallocha* Roxb.	Thymelaeaceae	Agarwood	Wood	Tree/Along with foothills & river valleys	CR/N	Essential oil, Agrol (0.08%)	2, 1, 7, 6 & 5	North South India (100-1,000 m altitude)
10.	*Arnebia benethamii* (Wall ex G. Don) John. Syn. *Macrotomia benthamii* (Wall) DC.	Boraginaceae	Gojaban	Leaves, Flower	Perennial herb/ Open slopes, stony or rocky places, forest of birch & rhododendron	CR/N	Arnebinus (0.37%), essential oil	4, 5 & 1	Kashmir to Uttaranchal (3,500-4,500 m altitude)
11.	*Atropa acuminata* Royle ex Lindl.	Solanaceae	Indian atropa	Leaves, roots	Perennial herb/In open slope	CR/N	Total alkaloids (0.81%)	4, 2 & 5	Kashmir to Himachal Pradesh (2,000-3,500 m altitude)
12.	*Baliospermum montanum* Muell.-Arg. Syn. *Jatropha Montana* Willd.	Euphorbiaceae	Danti	Seeds, Roots	Perennial herb/ Forest openings, edges & wastelands	LRnt/R	β-Sitosterol, resin & starch	4, 2, 1 & 5	Central India (upto 300 m altitude)
13.	*Berberis aristata* DC. Syn. *B. sikkimensis* (Schneid.) Arhndt.	Berberidaceae	Indian berberry	Bark of root	Spiny shrub/Open temperate forest of Himalaya	EN/R	Berbarine ($C_{20}0H_{18}O_4N$)	1, 2, 3 & 5	Indian North Western Himalayas (1,800-3,500 m altitude)
14.	*Berberis chitria* Lindl.	Berberidaceae			Spiny shrub/Sub-tropical to temperate forest	EN/N	Berbarine (5%, $C_{20}0H_{18}O_4N$)	1, 4, & 3	Kashmir to Kumaon regions of Uttaranchal (1,500-3,000 m altitude)
15.	*Berberis kashmirana* Ahrendt.	Berberidaceae	—	Bark of root	Shrub/Sub-alpine forest of Himalaya	CR	Berbarine ($C_{20}0H_{18}O_4N$)	2, 4, 1 & 5	J & K Himalayas (3,500-4,000 m altitude)
16.	*Berberis lycium* Royle.	Berberidaceae	Daruhaldi	Bark of root	Spiny shrub/In exposed arid places, sub-tropical to temperate	EN	Berbarine ($C_{20}0H_{18}O_4N$)	2, 1 & 5	J & K to Uttaranchal Himalayas (1,200-2,100 m altitude)

Contd...

Table Contd...

Sl.No.	Botanical Name	Family	English Name	Part Used	Habit & Habitat	IUCN Status	Active Constituents	Threats	Current Regional Distribution
17.	*Berberis petiolaris* Wall ex G. Don.	Berberidaceae	——	Bark of root	Large shrub/ Exposed slopes in temperate forest of Himalaya	CR	Berbarine ($C_{20}0H_{18}O_4N$)	1, 2, 3, 4 & 5	Yamuna Valley, Tehri Garhwal Himalayas (2,400-3,000 m altitude)
18.	*Bergenia ciliata* (Haw.) Sternb.	Saxifragaceae	Pasanbhed	Rhizome	Herb/Shady & moist rocks	VU/N	Bergenin (0.75%), gallic acid	1, 3, 4 & 5	Kashmir to Kumaon regions of Uttaranchal (1,200-2,500 m altitude)
19.	*Bunium persicum* Boiss Fedtcsh. Syn. *Carum bulbocastanum* acut. Non Koch.	Apiaceae	Black caraway	Fruit & Seeds	Perennial herb/On moist, open slopes.	EN/N	Essential oil (2.5%)	4, 8 & 9	Kashmir to Garhwal regions of Uttaranchal (2,000-3,500 m altitude)
20.	*Butea monosperma* (Witt.) Maheshwari Syn. *B. frondosa* Koe. ex Roxb.	Papilionaceae	Flame of the forest	Leaves, root & bark	Small tree/Dry deciduous forest	DD/R	Butin, Leucocinidin	4, 2, 1 & 5	Central India (upto 300 m altitude)
21.	*Celastrus paniculata* Willd. Syn. *C. Montana* Wight & Am.	Celastraceae	Black oil plant	Seeds, roots & bark	Woody liana/Sub-montane forest up to terai region	LRnt/R	Pristimerin	4, 2, 1 & 9	Central India (300-500 m altitude)
22.	*Cinnamomum tamala* (Ham.) Nees & Eberm.	Lauraceae	Indian Cassia	Leaves	Small tree/Sub-tropical regions	LRnt/R	Essential oil (0.3-0.6%)	2, 4 & 5	Indian North Western Himalayas (upto 1,000 m altitude)
23.	*Clerodendrum colebrokianum*	Verbenaceae	—	Leaves & root	Shrub/ Subtropical forest undergrowth along forest margins	VU/N		2, 3, 1 & 5	Northeastern India (800-1,500 m altitude)
24.	*Clerodendrum serratum* (L.) Moon	Verbenaceae	Bhargi	Leaves & root	Woody shrub/ Tropical moist deciduous forest	VU/R	Sapogenin	2, 4, 1 & 5	Central India (400-500 m altitude)
25.	*Coptis teeta* Wall.	Ranunculaceae	Gold thread	Rhizome	Herb/Moist, tropical, shady forest slopes	CR	Renin (0.5-2.7%) Berbatin (7.1-8.6%)	4, 2 & 5	Mishmi hills (2,500-3,000 m altitude)

Contd...

Table Contd...

Sl.No.	Botanical Name	Family	English Name	Part Used	Habit & Habitat	IUCN Status	Active Constituents	Threats	Current Regional Distribution
26.	*Cordia rothii* Roem & Schultz.	Ehretiaceae		Fruit & bark	Medium tree/Mixed dry deciduous forests & scrub forests	LRnt/R	Tanin	10, 2 & 5	Throughout Central India (upto 300 m altitude)
27.	*Costus lacerus*	Zingiberaceae		Bark	Shrub/Moist & marshy locations	DD/N	Not Known	—	Northwestern India (500-1,000 m altitude)
28.	*Crateriostigma plantagineum* Hochst.	Scrophulariaceae			Perennial herb/ Exposed, gravelly & shallow soils	CR/R		1, 2, 5 & 6	Western parts of M.P. (100-200 m altitude)
29.	*Curculigo orchioides* Gaertn. Syn. *C. malabarica* Wt.	Amaryllidaceae	Black Musli	Rhizome	Herb/Dry deciduous forest undergrowth in moist shady areas	VU/R		2, 4 & 5	Central, peninsular & western India. (500-700 m altitude)
30.	*Curcuma angustifolia* Roxb.	Zingiberaceae	East Indian Arrow	Rhizome	Herb/Dry deciduous forest	LRnt/R		4, 2, 3 & 5	Central India (500-800 m altitude)
31.	*Curcuma caesia* Roxb.	Zingiberaceae	Black Zedoary	Rhizome	Herb/Moist deciduous forest	CR/R		1, 4, & 5	Central India (500-800 m altitude)
32.	*Dactylorhiza hatagirea* D.Don Syn. *Orchis latifolia* auct. Non L.	Orchidaceae	Palm herbal medicine	Rhizome	Herb/Subalpine & Alpine Meadows	CR/R	Starch, Mucilage, volatile oil	4, 1 & 5	Kashmir to Kumaon region of Uttaranchal (2,500-4,000 m altitude)
33.	*Delphinium denudatum* Wall. ex Hook. f. & Thoms.	Ranunculaceae	Nirbisi	Roots	Perennial herb/ Open grasslands & margins of fields	CR/N	Ulphiuline	2	Kashmir to Kumaon region of Uttaranchal (2,500-3,000 m altitude)
34.	*Dioscorea deltoidea* Wall. ex Kunth.	Dioscoreaceae	Dioscorea	Rhizome	Herb/Open places	CR/R	Diosgenin (4.8-8.0%)	2, 4, 1 & 5	Indian Northwestern Himalayas. (1,500-2,500 m altitude)
35.	*Drymia indica* (Roxb.) Jessop. Syn. Urginea indica (Kunth.) Roxb.	Liliaceae			Herb/Rocky places along streams	VU/R		1, 3, 5 & 6	Central India (500-700 m altitude)
36.	*Evolvulus alsinoides* Linn. Syn. *E. hirtus* Lam.	Convolvulaceae	Sankhpuspi	Leaves & bark	Herb/Moist open fields	LRnt/R		2, 4, 1 & 5	Central India (upto 200 m altitude)

Contd...

Table Contd...

Sl.No.	Botanical Name	Family	English Name	Part Used	Habit & Habitat	IUCN Status	Active Constituents	Threats	Current Regional Distribution
37.	*Fritillaria roylei* Hook.	Ariacaeae		Bulb	Perennial herb/ Open alpine meadows	CR/N	Pimin ($C_{26}H_{43}O_3N$)	4, 2, 1 & 5	Kashmir to Kumaon region of Uttaranchal (3,000-3,500 m altitude)
38.	*Gastrochilus longiflora*	Zingiberaceae			Perennial stemless herb/On tree trunks in moist, broad-leaved subtropical forest	CR/N			Meghalaya, Mizoram, Arunachal Pradesh (1,000-1,500 m altitude)
39.	*Gentiana kurroo* Royle.	Gentianaceae	Indian Gantian	Rhizome	Herb/In Quercus forests, on bare hills & edges of rocks	CR/N	Gentianin	2, 4, 5 & 1	Kashmir to Kumaon region of Uttaranchal (2,000-4,000 m altitude)
40.	*Gloriosa superba* Linn.	Liliaceae	Glory lily	Roots	Herb/Along edges of moist tropical forest	EN/R	Gloriosine ($C_{22}H_{25}O_6N$), superbine, colchicine (1.3%)	4, 2, 1, 3 & 5	Tropical hills of Central India (upto 2,350 m altitude)
41.	*Gymnema sylvestre* (Retz) R. Br. Syn. *Asclepias germinata* Roxb.	Asclepiadaceae	Gurmar	Leaves	Climber/Deciduous forests	VU/R	Gymnemic acid	2, 1, 3 & 5	Central India (200-300 m altitude)
42.	*Hedychium coronarium* Koering	Zingiberaceae	Common Ginger lily	Rhizome	Perennial herb/ Moist tropical forest near stream & canals	EN/R	Essential oil	4, 1 & 5	Madhya Pradesh & Chhatisgarh (700-1,000 m altitude)
43.	*Hedychium spicatum* Ham. Ex Sm.	Zingiberaceae	Spiked Ginger	Rhizome	Perennial herb/ Moist & shady places	VU/N	Essential oil (4%)	4, 5 & 2	Himachal Pradesh to Kumaon region of Uttaranchal (1,500-2,500 m altitude)
44.	*Heracleum candicans* Wall. ex DC. Syn. *Heracleum lanatum* Mictux.	Apiaceae	Padara	Fruit	Perennial herb/ Along water courses, amongst stones & open slopes	EN/R	Xanthotoxin, coumarins	4, 5 & 1	Kashmir to Kumaon region of Uttaranchal (1,500-2,500 m altitude)
45.	*Hydnocarpus kurzii*	Flacourtiaceae	True Chaulmoogna	Seeds	Tree/Tropical evergreen forests	EN/N	—	1, 3 & 5	Northeastern India (200-800 m altitude)
46.	*Ilex khasiana*	Aquifoliaceae		Leaves & bark	Tree/Sun-tropical mixed evergreen forests	CR	lixine ($C_{53}H_{60}$, 2%)	2	Khasi hills (1,00-1,500 m altitude)

Contd...

Table Contd...

Sl.No.	Botanical Name	Family	English Name	Part Used	Habit & Habitat	IUCN Status	Active Constituents	Threats	Current Regional Distribution
47.	*Inula racemosa* Hook. f.	Asteraceae	Poshkar	Roots	Perennial herb/ Forest clearings & shrubberies	CR/N	Inulin, essential oil	1, 3 ,5 & 4	Kashmir to Kumaon region of Uttaranchal (2,000-3,200 m altitude)
48.	*Jurinea dolomiaea* Boiss. Syn. *J. macrocephala* (Royle) Clark.	Asteraceae	Doop	Roots	Perennial herb/ Grassland & open meadows in sub-alpine & alpine areas	LRnt	—	5, 11 & 2	Kashmir to Kumaon region of Uttaranchal (3,000-4,200 m altitude)
49.	*Lavatera cashmeriana* Cambess.	Malvaceae		Leaves	Perennial herb/ open grasslands, stony substrates	EN	—	4 & 5	Kashmir Himalaya (2,000-2,500 m altitude)
50.	*Luvunga scandens*	Rutaceae	Sugandh Kokila	Fruits	Shrub/Temperate, sacred forest	CR/N	Essential oil	2 & 3	Meghalaya (upto 1,500 m altitude)
51.	*Meconopsis aculeate*	Papaveraceae	Blue poppy	Roots	Herb/Rocky in alpine areas in between boundaries	EN/R	—	4, 2, 1 & 5	Indian Northeastern Himalayas (3,000 3,500 m altitude)
52.	*Nardostachys jatamansi.* DC. Syn. *Nardostachys grandiflora* DC	Valerianaceae	Spikenard	Rhizome	Perennial herb/ In rock crevices & moist shady places	CR/R	Essential oil	2, 4, 1 & 5	Uttaranchal (2,500 4,000 M altitude)
53.	*Nepenthes khasiana*	Nepenthaceae	Pitcher Plant	Leaves	Herb/Along road cutting & streams	CR	—	4, 2 & 5	Meghalaya (upto 500 m altitude)
54.	*Operculina turpethum* (L.) Silva. Syn. *Ipomea turpethum* R. Br.	Convolvulaceae	Nisoadh	Whole plant	Twining herb/Dry deciduous forest & fallow lands, wasteland	VU/R	Glucidine resin (10%)	1, 2, 4 & 5	Central India (upto 1,000 m altitude)
55.	*Paeonia emodi* Wall. ex Royle	Paeoniaceae	Himalayan Poney	Roots	Perennial herb/ Fringes of temperate forests	VU/N	Essential oil	1, 2, 4 & 5	Uttaranchal Himalayas (1,800-2,500 m altitude)
56.	*Panax pseudo ginseng* Linn.	Araliaceae	Asiatic Ginseng	Roots	Herb/Shady areas with an undergrowth of rhododendron forests along stream	CR/N	Panaqulon, Saponin (1%)	1, 2 & 5	Sikkim, Assam & Arunachal Pradesh (2,000-4,000 m altitude)
57.	*Picrirhiza kurroa* Royle ex Benth.	Scrophulariaceae	Kutki	Rhizome	Herb/On slopes & meadows of alpine Himalayas	EN/N	Picrorhizine, Kutkin (3.4%)	4, 2, 11 & 5	Indian Himalayas (2,500-4,000 m altitude)

Contd...

Table Contd...

Sl.No.	Botanical Name	Family	English Name	Part Used	Habit & Habitat	IUCN Status	Active Constituents	Threats	Current Regional Distribution
58.	*Podophyllum hexandrum* Royle. Syn. *P. emodi* wall ex Royle.	Podophyllaceae	Indian Podophyllum	Rhizome	Herb/Sub-alpine & alpine meadows	CR/N	Podophyllin	3, 4, 2, 1 & 5	Indian Himalayas (3,000-4,500 m altitude)
59.	*Polygonatum verticillatum* (L.) All.	Liliaceae	Meeta Dudhiya	Rhizome	Perennial herb/ Stony substrate, open places	EN/R	—	4, 5, 2 & 1	Kashmir to Kumaon region of Uttaranchal (2,000-3,500 m altitude)
60.	*Prezwalskia tangutica*	Solanaceae	Ma-niao-Pao	Roots	Perennial herb/ Alpine moraine areas	CR/N	Hyocine	11	Northern Sikkim (about 5,000 m altitude)
61.	*Rauvolfia serpentina* Benth.ex. Kurz. Syn. *Ophioxylon serpentina* L.	Apocynaceae	Rauvolfia	Roots	Herb/Moist to dry deciduous forest	EN/R	Reserpine	2, 4, 3, 1 & 5	Central India (100-300 m altitude)
62.	*Rheum austrate* D. Don Syn. *R, emodi* Wall. ex Meissn.	Polygonaceae	Indian Rhubard	Rhizome	Perennial herb/ Moist open grassy slopes & meadows	VU/R	Emodin, Chrysophanic acid	2, 4, 1 & 5	Kashmir to Kumaon region of Uttaranchal (3,000-4,000 m altitude)
63.	*Rheum nobile*	Polygonaceae		Rhizome	Perennial herb/ Open hilly slopes & screes	EN/N	Anthuqunon	2, 1 & 5	Sikkim, Assam & Arunachal Pradesh (3,500-4,500 m altitude)
64.	*Rhododendron anthopogon*	Ericaceae	Dhoop	Leaves	Shrub/Temperate to sub-alpine regions	VU/N	Anthroqunon	2 & 3	Sikkim, Assam, Darjeeling & Arunachal Pradesh (3,000-4,500 m altitude)
65.	*Rhus semialata*	Anacardiaceae	Dry galls	Fruit & Leaves	Tree/Subtropical semievergreen forest	VU/N	Tanin (50-80%)	2, 1 & 5	Northeastern India & foothills of Darjeeling (800-1,500 m altitude)
66.	*Saussurea costus* (Falc.) Lipsch. Syn. *S. lappa* (Decne.) Sch-Bip.	Asteraceae	Kuth	Roots	Perennial herb/ Open temperate to sub-alpine meadows	CR/N	Essential oil (1.5%), Saussurin	1, 2, 4 & 5	Kashmir to Garhwal region of Uttaranchal (2,000-3,000 m altitude)
67.	*Saussurea gossypiphora* D.Don	Asteraceae	Scared saussurea	Roots	Perennial herb/ Open places of alpine Himalayas	CR/R	—	4, 2 & 3	Kashmir to Kumaon region of Uttaranchal (3,500-5,700 m altitude)

Contd...

Table Contd...

Sl.No.	*Botanical Name*	*Family*	*English Name*	*Part Used*	*Habit & Habitat*	*IUCN Status*	*Active Constituents*	*Threats*	*Current Regional Distribution*
68.	*Saussurea obvallata* (DC.) Edgew.	Asteraceae	Bramkamal	Roots	Perennial herb/ Alpine zones-near glaciers	EN/R	—	4, 2, 1 & 5	Indian Northwestern Himalayas (3,000-5,100 m altitude)
69.	*Saussurea simpsoniana* (Field & Gard.)	Asteraceae	Doop	Roots	Woody herb/ Glacier zones of alpine Himalayas	EN/R	—	4, 2, 1 & 5	J&K, H.P. Uttaranchal (4,300-5,000 m altitude)
70.	*Swertia angustifolia* Buch.-Ham. ex D.Don. Syn. *Ophelia angustifolia* (Buch.-Ham. ex D.Don.) G. Don	Gentianaceae	Hilly Chirayata	Whole plant	Herb/Marshy areas frequently occurring in moist forests	EN/R	—	2, 4 & 5	M.P. & Chhatisgarh (400-700 m altitude)
71.	*Swertia chirayita* Roxb. Ex Flem.	Gentianaceae	Chiretta	Whole plant	Herb/Moist shady places	CR/R	Ophelic acid, chiratin	4	H.P. & Uttaranchal (1,500-3,000 m altitude)
72.	*Taxus wallichiana* Zuec. Syn. *T. baccata* Hook. f.	Taxaceae	Common Yew	Leaves & bark	Tree/In temperate mixed forests; occurring along with rhododendron, betula etc.	CR/R	Taxine ($C_{35}H_{49}O_{35-10}N$, 1.3%)	1 & 5	Sikkim, Assam & Arunachal Pradesh (1,500-3,000 m altitude)
73.	*Thalictrum foliolosum* DC.	Ranunculaceae	Yellow Zari	Rhizome	Perennial herb/ Open hill slopes and forests	VU/R	Bebrine (0.35%), Thalictrine (0.03%)	1, 2, 4 & 5	Uttaranchal (1,300-3,400 m altitude)
74.	*Tylophora indica* (Burm.f.) Merriel Syn. *Tylophora asthmatica* (L.f.) Wight & Am.	Asclepiadaceae	Emetic Swallow Wort	Leaves	Climber/Dry deciduous forests in the hills	VU/R	Tylophorine ($C_{24}H_{27}O_4N$)	1, 3 & 5	Central India (up to 1,000 m altitude)
75.	*Valeriana jatamansi* Syn. *V. wallichii*	Valerianaceae	Indian Valerian	Rhizome	Herb/Temperate Himalayas	CR/R	Volatile oil (0.5-2.12%)	2, 1, 4 & 5	Indian northeastern Himalaya (1,200-1,800 m altitude)

IUCN Status (Categories)

***Source*: BCPP Report, 1998**

CR = Critically endangered ; EN = Endangered ; VU = Vulnerable ; LR = Lower risk ; nt = near threatened : LR-lc = Least concern ; LR-cd- Conservation dependent: DD = Data deficient; NE = Not evaluated.

Threats*

1. Harvest for Medicine, 2. Loss of Habitat, 3. Human interference, 4. Overexploitation , 5. Trade , 6. Harvest for wood, 7. Edaphic factor, 8. Harvest for food, 9. Trade for parts, 10. Trampling , 11. Not Known.

Recommendations for Sustainable Management of Medicinal Plants

Central Government

(1) The Union Government should take all necessary steps including training of enforcement staff to initiate improved implementation of existing laws and regulations for medicinal plants conservation in India. (**Action**: MoEF, Customs and Excise Department and Controller of Drugs).

(2) Initiate and support medicinal plant trade studies based on which the plants list under schedules of WPA could be amended. This would also apply to the negative list under EXIM policy. (**Action**: MoEF, DISMH in the Ministry of Health, Central Council for Ayurveda, Unani and Siddha).

(3) Support research for development of cultivation techniques for important medicinal plants. Private initiatives for medicinal plants cultivation should also be encouraged by the government through liberal land laws and taxation relief. MoEF require to immediately put in place a certification/registration mechanism in place for non government medicinal plants nurseries wishing to raise specified plants (under WPA provisions) or species listed in the negative list of exports/CITES appendices. (**Action**: MoEF, DISMH. CIMAP. CDRI, NBPGR, ICAR, CSIR, and Ministry of Revenue and commerce).

(4) Develop a mechanism by which quality standards at the crucial supplier (wholesalers, retailers) level in large and small town in the country can be ensured.(**Action**: DISMH. ICAR, MoEF, Controller of Drugs).

(5) Sensitize the Customs and Excise department to the requirement of medicinal plants conservation and their crucial role in prevention of illegal trade.

State Government

(1) State governments which are the custodian of forest in their respective regions, need to devise and implement holistic policies and harmonious rules and regulations concerning collection, transport and processing of medicinal plants.

(2) The forest department in the state should be encouraged to complete useful and vast medicinal plant data currently buried in official files.

(3) State government may consider encouraging cultivation of medicinal plants in lands not otherwise being put to productive use.

(4) The laudable example of the Madhya Pradesh, where medicinal plants are being given due recognition and importance through establishment of a Non Timber Forest Produce Cooperative Federation may be emulated in other states.

(5) Awareness amongst sate forest department officials regarding their importance and crucial role as far as application of regulation under CITES and EXIM policy is concerned is conspicuous by absence. Even provisions of amended Wildlife (Protection) Act, 1972 as they apply to specified was found little appreciated.

(6) Forest/department/state government should look into the aspects of taxes/royalties collected from the medicinal herbs withdrawal/reduction in such taxes and royalties on medicinal herbs would help in a big way in increasing the incomes and well-being of the tribal population.

(7) Production of medicinal herbs should be taken up on a concrete business plan. The production should be linked and augmented wherever necessary based on actual need of a particular medicinal plants. There is an urgent need for undertaking proper demand survey based on data base information and co-relating the same for preparing effective production and processing plans.

(8) Efforts should be made to identify the different types of crude drug products and their demand by way of conducting comprehensive survey. This work can be entrusted to the various sate Forest Development Corporations in the respective States and at the national level by the agencies like TRIFED and NCDC.

Institutional responsibility for medicinal plants

(1) There is a plethora of organization dealing with various aspects of medicinal plants. There is a need for a nodal organization at national level to bring together various stakeholders and to deal with the entire range of issues pertaining to medicinal plants. It is hoped that the recently constituted Medicinal Plants Board would fulfill this role.

(2) In the absence of legal and policy provision pertaining to conservation, development and utilization, much unsustainable harvesting and other malpractices (adulteration, premature harvesting, use of substandard raw material, lack of scientific validation-certification etc.) are going on. The Medicinal Plants Board should frame appropriate legal and policy frameworks to address these problems. The Medicinal Plants Board will have to work closely with the proposed National Biodiversity Authority and the State Biodiversity Boards in this regard. The emphasis should be on facilitating sustainable harvests and management methods instead of merely regulating production and supply.

(3) It is general felt that, despite the impacts of the supply crisis, industries are not showing adequate concern towards conservation and sustainable supply of raw material and the well-being of collectors. It is recommended that the proposed nodal agency should bring together all interested parties i.e. the government agencies (Forest and Agriculture departments), industries, collectors/growers, NGOs etc. The nodal agency should ensure flow of information and develop participatory approaches for future development. It must be realized that it is in the interest of all stakeholders and the country that the science of ayurveda flourishes and its returns are shared by humanity at large, nationally and internationally.

(4) Local communities lack skill and tools to institute an operationally effective governance system to manage medicinal and aromatic plants.

(5) Role of Government organization and NGOs is more effective in facilitations, arbitration, conflict resolution, deal negotiations and plan development which is also acceptable to the formal administration and legal system.

(6) Small and homogenous communities are more successful in providing good governance and institutional sustenance.

Cultivation

(1) A comprehensive review regarding the current status of cultivation of medicinal plants in the country should be carried out. Existing bottlenecks being faced by farmers as well as companies should be identified and removed.

(2) Large-scale cultivation of important medicinal plants has been tried in different regions of the country. These practices have often used very heavy doses of inputs such as chemical fertilizers, pesticides etc. There is little information on the likely impact of these inputs on the medicinal plants properties of the plants. Until the results of further research on such impact become available, large-scale cultivation should be taken up with caution. It is recommended that the certification criteria of WHO (GAP: Good Agricultural Practices) be applied for the products from *ex-situ* sources.

Measures to control unsustainable harvests and improve production

(1) Unsustainable and destructive extraction has been widely reported. In the absence of full knowledge of sustainable harvesting regimes for different products, at least 50% of the fruit, roots, leaves, flowers and other materials collected should be left on the plants for future regeneration.

(2) There is a need to control premature harvesting of products such as fruits of *Emblica officinalis*. An initiative of this direction in Madhaya Pradesh has shown positive results.

(3) Value addition and processing at the collector level should be encouraged so that the collectors can earn the same amount of income from much less quantity.

(4) Direct link between the processing industry and primary collector/cultivator should be encouraged.

(5) Forest areas from where medicinal plants are extracted should be brought under sustainable forest management practices. Application of criteria and indicators, developed under the Bhopal-India Process, will help to tract progress towards sustainable forest management. In particular, the proposed 8 criteria and related 32 indicators of relevance to medicinal plants may be applied with suitable local modifications.

(6) Application of criteria and indicators in most likely to be effective if it is market-driven. It may be more realistic to start with a voluntary code of practice for the

production of ayurvedic medicines, complemented by education of both consumers and producers. However, it depends upon industry and / or consumer organizations to initiate this. The identification and ornaganisation of buyers who demand certified products would ultimately bring greater market access and higher prices for industry, and assured quality for consumers. The current thrust of the Planning Commission on increasing the exports of the ayurvedic formulations should be used as opportunity for exploring niche markets for certified ayurvedic products.

(7) Given that the collection and trade of medicinal plants are highly unorganized at the local levels, it is recommended that local level institutions such as FPCs need to be strengthened, so that the cultivation and conservation of medicinal plants may be integrated into the JFM arrangements. With proper micro-credit facilities and trading, micro-enterprises can be developed by these FPCs. A few initiatives in this regard have been already been started in some states.

(8) At present, export of prohibited plants (29 species including 18 mentioned in Appendix I and II of (CITES) is possible if these are present in some formulation (as given raw form) or if the label of the formulation does not mention the name of the species. This policy should be urgently reviewed.

Collection

(1) The primary collectors of medicinal herbs who inhabit or frequent the forest need to be educated and trained in identification, selection and in scientific collection of the forest produce. This is important form the point of view of obtaining produce of right quality at the right time maintaining at the same time health of the trees and of the forest. Organisations should be identified and programmes drawn for imparting training to the collectors of medicinal herbs in this regards.

(2) Collection of medicinal plants from the reserve forest and game sanctuaries are not permitted at present. There is a demand that collection should be allowed in such areas under the supervision of forest officials. The reason advanced is that tribal population who collect the produce from the forest should not be deprived of the benefits of this activity. It is felt that whereas economic exploitation of the forest wealth may be desirable, however, such activities should in no way endanger the endangered species of the animal for which such reserve forest sanctuaries are created.

(3) There is need for improvement of collection of medicinal plants. The traditional system of collection is far from satisfactory giving little attention to the various aspects like quality, maturity and availability of the product leading to ruthless collection

and total destruction of vegetation particularly of medicinal plants and herbs. It was recommended that modern scientific methods should be used in collection of medicinal herbs. The tribal people should be used in collection of medicinal plants. The tribal people should be educated and trained in these aspects. The concerned State Departments and agencies at state/national level can undertake this work.

(4) The training of persons/tribals involved in collection of medicinal plants and personnel of various cooperatives/market committees and other agencies would help in streamlining the marketing of these commodities. Training programmes should be arranged at grassroot level, preferably in forest areas/villages.

Grading/Processing & Value Addition

(1) Quality is the primary requirement for competing in the international market. The international market has become very specific and demanding in the matter of quality. Quality control through effective means in the only way to boost export. The effective means of quality control could be by self-regulation or by inspection and certification through inspection agencies.

(2) The grading at producer level prior to sale be introduced which would go a long way in realizing prices commensurate with the quality of the product. To facilitate grading, standards should be laid down for crude raw material of medicinal plants. The market committees and co-operatives/corporations should provide necessary infrastructure in this regard.

Establishment of Industry

(1) It is felt that it will be more economical to sell forest products in processed/value-added form. It is therefore, recommended to explore the feasibility of establishing processing units at appropriate places, which may help in achieving this object.

Marketing Channels

(1) Margin of collectors & glowers is higher in trading of legal products.

(2) Quality does matter to traders & end-users and fetches higher price.

(3) Traders have no or less interest in sustainable harvesting of medicinal & aromatic plants.

(4) Traders in higher end of chain have full knowledge of market trends & prices and do not share the information readily.

(5) Tribal and other co-operative should be financially strengthened to enable them to take active participation in the marketing of medicinal plants. Their effective presence in the market will create competition and will also influence the prices paid to the primary producers.

(6) It was felt that there are a number of lacunae in the present marketing of respect of medicinal plants. In majority of markets, these commodities have not been notified for the purpose of regulation, perhaps, in view of their less quantum of arrivals. Further, the importance of medicinal herbs differs from state to state.

(7) The market intelligence and marketing extension services should be strengthened for effective disposal of forest produce.

(8) The co-operative should function in a way to eliminate the intermediates to the extent possible, so that the tribal get remunerative prices for the produce.

(9) Marketing finance on easy terms may be made available to the tribal to free them form the clutches of money lenders/contractors etc.

(10) Commodity specific studies may be conducted which can be used as renewable source of energy like Raton Jot etc.

(11) There are multiplication of agencies in the marketing of medicinal plants. Their areas of activities must be demarcated.

(12) There should be inter-linkages in institutions/organizations in coordination of different programmes so that the marketing efficiency is increased.

(13) The monopoly procurement operation is in vogue in Maharashtra. There should be an option to the collector for selling medicinal plants in the open market on competitive prices and the suitable infrastructure facilities, therefore, in primary tribal markets need to be developed.

Institutional & Technological Innovations

(1) Local organizations and enterprise need to be made better aware of the needs of traditional modern drug producers and skilled in extraction, processing and marketing of medicinal and aromatic plants.

(2) Industry should enter into buy-back arrangement with local community groups and individual producers.

(3) Market and price information should be made available by industry to the local traders and producers and collectors.

Micro & Macro Policies

(1) Impact of macro policies on conservation & development is slow and that too only if enforces strictly and in an impartial manner.

(2) Micro policies impact directly and immediately, as enforcement is quicker and relatively enforced even-handedly due to social pressure.

(3) Policies that provide incentives to maximize production, processing and marketing to enterprises and individual families may help conservation of medicinal plants as alternate resources are created.

Glossary of Medical Terms

Abortifacient : Drug or agent inducing expulsion of a non-viable fetus.

Abscess : Localized collection of plus formed as a reaction to pyegenic organisms.

Acne : A skin condition common in adolescence, in which blackheads (comedones) are associated with a popular and pustular eruption of the pilosebaceous follicles.

Acute : Short and severe, not long drawn out or chronic.

Affection : Any diseases or pathlogical condition.

Ague : Malarial fever with successive stages of fiver and chills.

Alopecia : Baldness which can be congential, premature or senile.

Alterative : Able to restore health.

Amenorrhoea : Absence of the menses.

Amoebiasis : Infection of large intestine by the protozoon *Entamoeba histolytica,* which it causes ulceration by invasion of the mucosa.

Anaemia : A disorder due to a deficiency in the number of red blood cells or of their haemoglobin content, or of both.

Analgesic : A drug which relieves pain.

Anasarca : Serous infilteration of the cellular tissues and serous cavities; generalized oedema.

Anodyne : A remedy which relieves pain.

Anthelmintic : Any remedy for the destruction or elimination of intestinal worms.

Antidote : A remedy which counteracts or neutralizes the action of a person.

Antifebrile : Any agent which reduces or allays fever.

Anti-inflammatory : Any agent which prevents inflammation.

Antiperiodic : Any agent which prevents the periodic returns of a disease.

Antipyretic : Any agent which allays or reduce fever.

Antiscoebutic : Any agent which prevents or cure scurvy.

Antispasmodic : Any measure used to relieve spasm occurring in muscle.

Aperients : Drugs which stimulate evacuation of the bowel.

Aphrodisiac : An agent which stimulates sexual excitement.

Appetizer : A small amount of food or drink taken to stimulate the appetite.

Arthrorsis : An articulation or joint.

Asthma : Paraxysmal attack of difficulty in breathing (disease of lungs causing difficulty in breathing).

Astringent : An agent which contracts organic tissue.

Beri-beri : A deficiency diseases caused by lack of aneurine (Vitamin B1).

Bilharziasis : Infestation of the human body by schistosoma (Blood flukes) from drinking, or bathing in infected water.

Bitters : Substrates, the extracts of which are used as stomachics.

Boil : A furncle. A acute inflammatory condition, surrounding a hair follicle; caused by the Staphylococcus aureus.

Bowel : An intestine.

Cardialgia : Pain in the heart.

Carminative : Having the power to relieve flatulence and associated colic.

Cataract : An opacity of the crystalline lens or its capsule. Eye complaint of old age.

Catarrh : Inflammation of a mucous membrane with constant flow of mucus.

Cathartic : Purgative.

Cephalagia : Pain in the head.

Cholagogue : A drug which causes and increased flow of bile into the intestine.

Chronic : Lingering, lasting, opposed to acute.

Colic : Severe pain resulting from periodic spasm in an abdominal organ.

Conception : The act of becoming pregnant by the impregnation of the ovum by the spermatozoon.

Consumption : A wasting away of the tissues of the body, in tuberculosis of the lungs.

Contagion : Communication of diseases from body to body.

Contraceptive : An agent used to prevent conception.

Convulsions : Involuntary contractions of muscles.

Corn : A cone-shaped, overgrowth and hardening of epidermis.

Cramp : A painful involuntary contraction of a muscle.

Cystitis : Inflammation of the urinary bladder.

Demulcent : A slippery, mucilaginous fluid which allays irritation and soothes inflammation, especially of mucous membranes.

Dentifrice : Paste or powder, for use in cleaning the teeth.

Depressant : Able to reduce nervous or functional activity.

Diaphoretic : Increased secretion of urine.

Diuretic : An agent which increases the flow of urine.

Dyspepsia : Indigestion.

Dystrophy : Defective nutrition.

Dysuria : Difficult or painful micturtion.

Ecbolic : Any agent which stimulates contraction of the gravid uterus and hastens expulsion of its contents.

Emaciation : Excessive leanness, or wasting of body tissue.

Emetic : medication which stimulates vomiting.

Emmenagogue : Drug used to bring on menstruation.

Emollient : An agent which softens and sooths skin or mucous membrane.

Epilepsy : The 'fit' is caused by an abnormal electrical discharge that disturbs cerebration and usually results in loss of consciousness.

Expectorant : A drug which promotes or increase expectoration.

Febrifuge : Any drug or agent for reducing fever.

Flatulence : Gastric and intestinal distension with gas.

Galactagogue : An agent inducing or increasing the flow of milk.

Gargle : The act of washing the throat.

Gastralgia : Pain in the stomach.

Germicide : An agent which kills germs.

Gonorrhea : An infectious disease of venereal origin in adults.

Haemoptysis : Blood in the urine.

Haemorrhoids : The condition resulting from infestation with worms.

Hemicrania : Unilateral headache as in migrain.

Hemicrania : Paralysis of one side of the body.

Hypnotic : A drug which produces asleep resembling natural sleep.

Hypotension : Abnormally low blood pressure.

Hypothermia : Below normal body temperature.

Hysteria : A mental disorder characterized by emotional outbursts, and often physical symptoms such as paralysis.

Inflammation : The reaction of living tissues to injury, infection, or irritation; characterized by pain, swelling, redness and heat.

Influenza : An acute viral infection of the nasopharynx and respiratory tract.

Insanity : Relatively permanent disorder of the mind.

Jaundice : A condition characterized by a raised bilirubin level in the blood.

Laxative : A mild aperient.

Leucorrhoea : A sticky, whitish vaginal discharge.

Leukaemia : Blood disease in which the white cells are abnormal in type or number.

Malignant : Virulent and dangerous.

Menorrhagia : An excessive regular menstrual flow.

Metorrhagia : Uterine bleeding between the menstrual periods.

Narcotic : A drug which produces abnormally deep sleep.

Nausea : A feeling of sickness without actual vomiting.

Neuralgia : Pain in the distribution of a nerve.

Oedema : Dropsy. Abnormal infiltration of tissues with fluid.

Ophthalmia : Inflammation of the testicle.

Otitis : Inflammation of ear.

Ozaena : Atrophic condition of the external auditory meatus.

Poultice : A local moist and often heated application for the skin used to improve the circulation, treat inflamed areas etc.

Refrigerant : Increasing and decreasing at periodic intervals.

Rubefacients : Substances which, when applied to the skin, cause redness.

Sedative : An agent which disease caused by lack of vitamin C.

Sialagogue : A agent which induces profound sleep.

Stomachics : Agents which increase the appetite.

Tonic : A medical preparation that improves the functioning of the body or increases feeling of well-being.

Tympanitis : Inflammation of the tympanum.

Ulcer : An open sore in a body surface.

Urticaria : An allergic skin eruption characterized by multiple, circumscribed, smooth, raised, pinkish, itchy weals, developing very suddenly, usually lasting a few days, and leaving no visible trace.

Vermifuge : An agent that expels intestinal worms.

Wart : Any firm abnormal elevation of the skin caused by a virus.

Bibliography

Anon (1945-1976) Wealth of India, Raw Material, Vol. I-XI, Publication and Information Directorate, C.S.I.R., New Delhi.

Anon (1948) Report on the committee on Indigenous system of medicine, Vol. 1, Ministry of Health, Govt. of India, New Delhi.

Anon (1986) The useful plants of India. Publication and Information Directorate, C.S.I.R., New Delhi.

Anon (1998) Market trends in production, price, export, import etc., Jour. Med. Aro. Pl. Sci., Vol. 20, pp 106-112.

Anon (1999) Cites listed Medicinal Plants of India.

Anon (2000) Report of the Task force on conservation & sustainable use of medicinal plants. Planning Commission (Government of India), New Delhi.

Arias-Castro, C.; Scragg, A.H. & Rodriguez, M. (1993) The effect of cultural condition on the accumulation of formation by suspension cultures of *Glycyrrhiza glabra. Plant Cell Tissue & Org. Culture* vol 34: pp 63-70.

Arora, R.K. (1993). Himalayan plants resources: Diversity and Conservation. In; Himalayan Bio-diversity Conservation Strategies (Ed. U. Dhar). Gyanodaya Prakashan, Nainital. pp. 39-44.

Atal, C.K. & Kapoor B. M. (1989) Cultivation and Utilization of Medicinal Plants. RRL, Jammu-Tawi, J & K.

Atkinson, E.T. (1980) Flora of the Himalayas (with special reference to Kumaon, Garhwal and parts of Tibet; Cosmo Publication, New Delhi.

Atul, C.K. & Kapur, B. M. (1977) Cultivation and Utilization of medicinal and Aromatic plants. Regional Research Laboratory, Jammu-Tawi.

Awashti, R.K & Rana, R.S. (2001) Large Cardamom: Production and Forest Conservation in Sikkim. WWF-India. New Delhi.

Bajaj, Y.P.S. (1990) *In vitro* Production of haploids and their use in cell genetics and plant breeding. In Biotechnology in Agriculture and forestry 12 (ed.), Medicinal and Aromatic Plants 11, Springer-Verlag, Berlin.

Bajaj,Y.P.S. & Simola, L.K. (1991) *Atropa belladona* L. *In vitro* culture, regeneration of plants, cryopreservation and the production of tropane alkaloids. In Biotechnology in Agriculture and forestry 15, Medicinal and Aromatic Plants 111, Springer- Verlag, Berlin.

Banerjee, R. C. (1994). Standardization and grading of Minor forest Products. Agricultural Marketing. Apr-Jun issue; pp 24-27.

BCCP (1998) Biodiversity Conservation Prioritization Project (BCCP) India, Zoo Outreach Organisation/ BSG, India.

Bhattacherjee, S. K. (2001) Hand book of Medicinal Plants. 3rd Ed. Pointer Publishers, Jaipur, India.

Bhattarai, N.K. (1989) Traditional phytotherapy among the Sherpas of Helambu, Central Nepal. *Jour. Ethenopharmacology*, Vol. 27 (1&2): pp. 45-54.

Biwas, K (1956) Common Medicinal Plants of Darjeeling and Sikkim Himalayas, Government Printing Press, West Bengal.

Brandis, D (1874) Forest flora of North West and Central India. Allen & Co. London.

Chadha, K.L. & Gupta, R. (1995) Advances in Horticulture, Medicinal and Aromatic plants, Vol.11, Malhotra Publishing House, New Delhi.

Chopra, R. N., Nayar, S. L. & Chopra, I. C. (1974). Supplement to the glossary of Indian Medicinal Plants, C.S.I.R. Publication, New Delhi.

Chowdhary, H. J. and Wadhava, B. M. (1984). Flora of Himachal Pradesh Analysis. Vol. 1-3, Botanical Survey of India, Kolkatta (W.B).

Curtis, J. T. and Cotton, G (1956) Plant ecology work book. Laboratory field reference manual. Burger's Publishing Co., Minnesota, USA.

Dastur, J. F. (1951). Useful Plants of India and Pakistan, T. Sons & Co. Ltd. Bombay.

Dey, A.C. (1964). Indian Medicinal Plants Used in Ayurvedic Preparations, Bishen Singh Mohendra Pal Singh, Dehradun.

Drury, C. H. (1978 Reprint) The Useful Plants of India, International Book Distributors, Dehradun.

Edwards, D.M. and Bowen, M. R. (1993) Focus on Juri buti. FRSC, Kathmandu, Nepal.

Elhag, H.M. ; El- Domiaty, M.M.; El Feraly, F.S.; Mossa, J.S. & El-Olemy,M.M. (1992). Selection and micropropagation of high artimisin in producing clones of *Artimisia annua* L. *Phytotherapy Res.* Vol. 6: pp 20-24.

Fransworth, N.R. and Morris, R.W. (1976) Higher plants: the sleeping giant of drug development. *Am. J. Pharm. Edu.* Vol. 148, pp. 46-52.

FRIS Project Paper 4. Forest Research Information System Project, Ministry of Forest and Soil Conservation, Kathmandu, Nepal. p. 27.

FRLHT (1996)" Conserving a National Resource: Need for a National Policy and National Programme on Medicinal plant conservation". Unpublished background paper for National Consultation on formulating a medicinal plant conservation policy in India. Med plant Conservatory Society, Bangalore.

Gautam, P.L. ; Singh, B.M. & Sharma, N. (1998). Bioprospecting and conservation of Non-Timber Forest Products *vis a vis* Medicinal and Aromatic Plants, Some approaches to forest Management. *Forest Usufructs*, Vol.1 (1&2) pp 143-150.

Grieve, M. (1989). A Modern Herbal, Vol I& II. Hafner Publishing Co., New York.

Gupta, R. K. (1968). Cultivation of *Digitis lanata*. In Chamba Hills, Himachal Pradesh. Indian Forester, Vol. 93(1), 33-40.

IUCN, WHO, & WWF (1993) Guidelines on the Conservation of Medicinal Plants. International Union for the Conservation of Nature and Natural Resources, Gland, Switerland.

Jain, S. K. (1968) Medicinal Plants. National Book Trust, New Delhi.

Karki, M. (1998). A review of the medicinal plants Sector in India. IDRC, Canada.

Kaul, M.K. (1996). Strategies for sustainable use of threatened medicinal plant resources in Western Himalaya 1n : Ethnobiology in Human Welfare, (Ed. S.K. Jain). 284-286, Deep Publisher, New Delhi.

Kirtikar, K. R. and Basu, B. D. (1975) Indian Medicinal Plants. Vol. I-IV. Allahabad, India.

Kuipers, S. E. (1997) Trade in Medicinal plants. In: Medicinal plants for forest Conservation and Health care. NWFP Series No. 11. FAO, Rome.

Maheshwari *et al* (1988) Studied Economics of Intercropping of *Rauwolfia serpentina* with other crops such as soya bean, onion, garlic, etc.

Maheshwari, P. and Singh, U. (1981). Dictionary of Economic plants in India. ICAR, New Delhi.

Maheshwari, S.K.; Dhahtonde, B.N. & Gangrade, S.K. (1988) Differential response of intercropping of *Rauvolfia* with other crops. Progress in Ecology, Symp. Proc. (Today and Tomorrow Printers and Publishers, New Delhi) 10 : pp 121-126.

Malla, S.B., Shakya, P.R.; Rajbhandari, K.R.; Bhattarai, N.K.; Subedi, M.N (1995). Minor Forest Products of Nepal: General status and trade.

May, R.M. (1988). How many Species are there on earth? *Science*, Vol. 20, September, 16.

Menon, A. K. (1960). Indian Essential oil (A review). C.SI.R., New Delhi.

Mishra, R (1968) Ecology work book. Oxford and IBH Publishers, New Delhi.

Muller-Dombois, D and Ellenberg, H. (1974) Aims and methods of vegetation ecology. John Wiley, New York.

Natural Medicinae Marketing (1996) Market Reports-Traditional Chinese Medicines. The Chinese Market and International opportunities, London, UK.

Postal, S. (1994) Carrying Capacity: Earth's Bottom Line. State of the World, World Watch Institute, Washington, D.C.

Ramp, D.D. (1986) Biological extinction in Earth History. *Sciences*, Vol.28.

Rawat, T. (1997) Growing market of herbal medicines. Jour. NTFP. Vol. 4 (1/2), pp 87-88.

Samuel, J. C.; Sivaraman, K and Singh, H. P. (2001) Medicinal and Aromatic plants. South Asian. Agri Business & Horticulture. Vol. 1 (4): 49-54.

Shah, R.R & Dalal, K.C. (1980) *In vitro* multiplication of *Glycyrrhiza. Curr. Sci.* Vol. 49: pp 69.

Sharma, N. & Chandel, K.P.S. (1992) Low temperature storage of *Rauvolfia serpentina,* An endangered, endemic medicinal plant. Plant Cell Reports Vol. 11: pp 200-203.

Sharma, P. (1996) Opportunities in and consultants to the sustainable use of Non Timber Forest resources in the Himalayas. ICIMOD,. Vol. 25, pp 9-11.

Sharma, R (1999) *Artemisia maritima*. MFP News, Vol. 9, No. 3 p. 13-14.

Sharma, R (1999) *Colchicum lutem.* MFP News, Vol. 9, No. 2 p. 13.

Sharma, R (1999) *Gentiana kurroo*. MFP News, Vol. 9, No. 4 p. 14.

Sharma, R (1999) *Taxus baccata*. MFP News, Vol. 9, No. 1 p. 17-18.

Sharma, R (2000) *Curcuma caesia.* MFP News, Vol. 10, No. 4 p. 14

Sharma, R (2000) *Gloriosa superba.* MFP News, Vol. 10, No. 3 p. 10-11.

Sharma, R (2000) *Saussurea costus*. MFP News, Vol. 10, No. 2 p. 17-18

Sharma, R (2000) *Valeriana jatamansi*. MFP News, Vol. 10, No. 1 p. 14.

Sharma, R (2001) *Poenia emodi*. MFP News, Vol. 11, No. 1 p. 14

Sharma, R (2001) *Tylophora indica*. MFP News, Vol. 11, No. 2 p. 10

Sharma, R (2002) Aushadhaya avum Sugandhiya Paudhuon ki Krishi Taknik. Daya Publishing House, New Delhi.

Shiva, A & Sharma, A (2000). Annonated bibliography on Non-timber Forest Products (NTFP) : Related articles. COMFORPT, Dehradun.

Snedecor, G.W and Cochran, W.G. (1972) Statistical Methods. Iowa State University Press, USA.

Srivastava, G. N. (1994) Medicinal and Aromatic Plants - A potential source of forest produce. Agricultural Marketing. Apr-Jun issue; pp 50-54.

Srivastava, J.; Lambert J. & Vietmeyer, N. (1995) Medicinal Plants; A growing role in development, Washington D.C., USA, Agricultural and Natural Resources Department, The World Bank.

Subrat, N.; Iyar, M & Prasad, R (2002) Ayurvedic medicine Industry: Current Status and sustainability. ETS Publishers, New Delhi.

Taylor, D. (1997) Herbal Medicine. *Environmental Health Perspectives* Vol. 105, No.11

Uniyal, M.R. (1989). Medicinal flora of Garhwal Himalaya. Shree Baidyanath. Ayurveic Bhawan Pvt. Ltd. Nagpur.

Wadia, D.N. (1953). Geology of India. Mac. Millan & Co. Ltd. London.

Index

D

N

O

P

Q

R

S

T

U

V

W

X

Z